Lecture Notes in Control and Information Sciences

Edited by M. Thoma and A. Wyner

142

To Nirmal
with best regards,
Charlie

A. N. Gündeș, C. A. Desoer

Algebraic Theory of Linear Feedback Systems with Full and Decentralized Compensators

Springer-Verlag
Berlin Heidelberg New York
London Paris Tokyo Hong Kong

Authors
A. Nazli Gündeş
Dept. of Electrical Engineering
and Computer Science
University of California
Davis, CA 95616
USA

Charles A. Desoer
Dept. of Electrical Engineering
and Computer Sciences
University of California
Berkeley, CA 94720
USA

ISBN 3-540-52476-2 Springer-Verlag Berlin Heidelberg New York
ISBN 0-387-52476-2 Springer-Verlag New York Berlin Heidelberg

Printing: Mercedes-Druck, Berlin
Binding: B. Helm, Berlin
2161/3020-543210 Printed on acid-free paper.

PREFACE

The algebraic theory of linear, time-invariant, multiinput-multioutput feedback systems has developed rapidly during the past decade. The factorization approach is simple and elegant; it is suitable for both continuous-time and discrete-time lumped-parameter system models and many of its results apply directly to distributed-parameter systems. Major achievements of this algebraic theory are: 1) several equivalent formulations of necessary and sufficient conditions for stability, where the concept of stability is defined with flexibility to suit various applications; 2) the parametrization of all achievable stable input-output maps for a given plant; 3) the parametrization of all compensators that stabilize a given plant; and 4) a general method which is suitable to study different system configurations.

In this volume we aim to unify the algebraic theory of full feedback and decentralized feedback control systems. Our main focus is the parametrization of all stabilizing compensators and achievable stable input-output maps for three particular feedback configurations: the standard unity-feedback system; the general feedback system in which the plant and the compensator each have two (vector-) inputs and two (vector-) outputs; and the decentralized control system in which the compensator is constrained to have a block-diagonal structure.

Several of the results we present are well-known to control theorists. We clarify and unify the presentation of these results, remove unnecessary assumptions and streamline the proofs. Among the new developments in this volume are a characterization of all plants that can be stabilized by decentralized feedback and the parametrization of all decentralized stabilizing compensators. The introduction to each chapter includes a list of the important results.

A good preparation for the material in this volume is a graduate-level course in linear system theory and some familiarity with elementary ring theory.

We gratefully acknowledge the support of the Electrical Engineering and Computer Science Departments of the University of California at Davis and Berkeley and the National Science Foundation (Grant ECS 8500993). We thank Jackie Desoer and Güntekin Kabuli for their patience.

Table of Contents

Chapter 1

INTRODUCTION

In this volume we present a unified algebraic approach to the study of linear, time-invariant (lti), multiinput-multioutput (MIMO), full-feedback and decentralized feedback control systems. This approach applies to continuous-time as well as discrete-time, lumped-parameter system models. Much of this theory applies directly to distributed-parameter systems (see e.g. the books [Blo.1, Cal.3, Fei.1, Vid.1] and the papers [Cal.1, Cal.2]).

We use a factorization approach which is based on elementary ring theory. In order to separate algebra from control theory, we collect all relevant purely algebraic facts and theorems in Chapter Two. We do not include basic definitions and properties of rings; elementary ring definitions (entire ring, principal ring, ideal of a ring, ring of fractions, etc.) can be found in several texts in algebra [Bou.1, Coh.1, Jac.1, Lan.1, Mac.1]-- we recommend the brief review in [Vid.1, Appendix A, B]. The goal of Chapter Two is to present the conceptual tools and key results needed for the systematic use of right- left- and bicoprime factorizations.

In Chapter Three we study two classes of feedback systems that have no restrictions on the compensator structure. In Chapter Four we consider the restriction that the compensator transfer-function is a block-diagonal matrix.

The main issues that we address in Chapters Three and Four are: closed-loop stability, the parametrization of all stabilizing compensators and of all achievable stable closed-loop input-output (I/O) maps. Three particular feedback configurations are considered: The first one is the unity-feedback system, which we call $\mathbf{S}(P, C)$; this is the standard multiinput-multioutput feedback system made up of a plant P and a compensator C, where there are no restrictions on the compensator structure. The second configuration, which we call $\Sigma(\hat{P}, \hat{C})$, is a more general interconnection where the plant and the compensator have inputs and outputs that are not utilized in the feedback-loop; the unity-feedback system $\mathbf{S}(P, C)$ is a special case of $\Sigma(\hat{P}, \hat{C})$. The third configuration we consider is the decentralized control structure; this is a special case of the unity-feedback system $\mathbf{S}(P, C)$, where only certain outputs are available to be fed back to certain inputs; in this case the compensator is constrained to have a block-diagonal structure. The two-channel decentralized control system, which we study in detail, is called $\mathbf{S}(P, C_d)$; the results are extended to the multi-channel (m–channel) decentralized control system $\mathbf{S}(P, C_d)_m$.

Section 3.2 focuses on the unity-feedback system $\mathbf{S}(P, C)$ (Figure 3.1). It is well-known that, for any MIMO plant P, there exists a dynamic feedback compensator C such that the closed-loop system $\mathbf{S}(P, C)$ is internally stable. The parametrization of all stabilizing compensators based on a right-coprime factorization $N_p D_p^{-1}$, a left-coprime factorization $\tilde{D}_p^{-1} \tilde{N}_p$, or a bicoprime factorization $N_{pr} D^{-1} N_{pl} + G$ of the plant P is a fundamental tool in describing the achievable closed-loop performance of $\mathbf{S}(P, C)$.

In Section 3.3 we consider the general feedback system $\Sigma(\hat{P}, \hat{C})$ (Figure 3.9). In this system, the plant $\hat{P}$ and the compensator $\hat{C}$ both have two (vector-) inputs and two (vector-) outputs. This configuration takes into account such cases where the regulated plant-output is not necessarily the same as the measured output or where the plant is directly affected by exogenous disturbances.

In Sections 4.2 and 4.3 we study the two-channel decentralized control system $\mathbf{S}(P, C_d)$ (Figure 4.1). We characterize the class of all plants that can be stabilized by decentralized output-feedback and parametrize all decentralized stabilizing compensators. In Section 4.4 we consider systems that have rational transfer functions and we study the relationship between decentralized stabilizability and decentralized fixed-eigenvalues. In Section 4.5 we extend the results of Sections 4.2, 4.3 and 4.4 to the multi-channel decentralized control systems $\mathbf{S}(P, C_d)_m$ (Figure 4.7).

We assume throughout that a transfer-function approach makes sense; in particular, the plant and the compensator subsystems in the feedback-system have no hidden-modes associated with unstable eigenvalues so that their transfer functions describe the behavior of these subsystems adequately for stability purposes.

The style and the results in Chapters Two and Three are inspired by [Vid.1, Net.1, Des.3]; some results are based on our original work (e.g. [Des.4, 5, 6, 7]). Vidyasagar's book [Vid.1] has a list of references for previous related work. There is also a considerable amount of previous work on the existence issues for decentralized compensators and on various formulations of decentralized fixed-modes (e.g. [And.1, 2, Dav.1, 2, Fes.1, Tar.1, Vid.4, Wan.1, Xie.1]. The characterization of all plants stabilizable by decentralized feedback and the parametrization of all decentralized compensators that is presented in Chapter Four is a new development.

Chapter 2

ALGEBRAIC FRAMEWORK

2.1 INTRODUCTION

This chapter gives an integrated development of the algebraic facts that are used in the following chapters. The reader is assumed to be familiar with basic properties of rings; the material presented in [Vid.1, Appendix A, B] provides sufficient background. More detailed discussions on ring theoretic concepts can be found in [Bou.1, Coh.1, Jac.1 (Sections 2.1-2.3, 3.7), Lan.1, Mac.1].

Notation

H is a principal ideal domain.

$\mathrm{J} \subset \mathrm{H}$ is the group of units of H.

$\mathrm{I} \subset \mathrm{H}$ is a multiplicative subset of H, where $0 \notin \mathrm{I}$, $1 \in \mathrm{I}$.

$\mathrm{G} := \mathrm{H}/\mathrm{I} = \{ n/d \mid n \in \mathrm{H},\ d \in \mathrm{I} \}$ is the ring of fractions of H with respect to I.

$\mathrm{F} := \mathrm{H}/(\mathrm{H}\backslash 0) = \{ n/d \mid n, d \in \mathrm{H},\ d \neq 0 \}$ is the field of fractions of H.

$\mathrm{G}_S := \{ x \in \mathrm{G} \mid (1+xy)^{-1} \in \mathrm{G}, \text{ for all } y \in \mathrm{G} \}$ is the (Jacobson) radical of G.

□

Note that: **(i)** every element of I is a unit of G; **(ii)** $\mathrm{F} \supset \mathrm{G} \supset \mathrm{H} \supset \mathrm{I} \supset \mathrm{J}$; **(iii)** if $x \in \mathrm{I}$ then $x \notin \mathrm{G}_S$.

Let $a, b \in \mathrm{H}$; then a and b are *associates* (denoted by $a \sim b$) iff there exists $u \in \mathrm{J}$ such that $a = b\,u$; note that $a \sim 1$ iff $a \in \mathrm{J}$. " $\sim$ " is an equivalence relation on H.

The set of matrices with elements in H is denoted by $\mathfrak{m}(\mathrm{H})$; this notation is used when the actual order of the matrices is unimportant. Where it is important to display the order of a matrix explicitly, a notation of the form $A \in \mathrm{H}^{n_o \times n_i}$, $B \in \mathrm{H}^{n_i \times n_i}$ is used instead of A, $B \in \mathfrak{m}(\mathrm{H})$. In the study of linear control systems, the set $\mathfrak{m}(\mathrm{H})$ corresponds to stable systems; therefore we call a matrix H–stable iff $A \in \mathfrak{m}(\mathrm{H})$.

The identity matrix is denoted by I ; in some cases the order of the identity matrix is indicated with a subscript as in I_n .

Let $A \in \mathfrak{m}(\mathrm{H})$; then A is called H–unimodular (G–unimodular) iff A has an inverse in $\mathfrak{m}(\mathrm{H})$ ($\mathfrak{m}(\mathrm{G})$, respectively); equivalently, A is H–unimodular (G–unimodular) iff $\det A \in \mathrm{J}$ ($\det A \in \mathrm{I}$, respectively).

Let $X \in \mathfrak{m}(\mathrm{G}_s)$, $Y \in \mathfrak{m}(\mathrm{G})$ and $Z \in \mathfrak{m}(\mathrm{G})$ have appropriate dimensions so that XY and ZX are defined; then $XY \in \mathfrak{m}(\mathrm{G}_s)$, $ZX \in \mathfrak{m}(\mathrm{G}_s)$, $(I + XY)^{-1} \in \mathfrak{m}(\mathrm{G})$ and $(I + ZX)^{-1} \in \mathfrak{m}(\mathrm{G})$.

2.2 PROPER STABLE RATIONAL FUNCTIONS

Let $\mathbb{C}_+ := \{ s \mid \mathrm{Re}\, s \geq 0 \}$ denote the closed right-half-plane and let $\mathbb{C}_- := \{ s \mid \mathrm{Re}\, s < 0 \}$ denote the open left-half-plane of the field $\mathbb{C}$ of complex numbers. Let $\mathfrak{U}$ be a nonempty subset of $\mathbb{C}$ such that $\mathfrak{U}$ is closed and symmetric about the real axis, and $\mathbb{C} \setminus \mathfrak{U} =: \mathbb{D}$ is nonempty; $\mathbb{D}$ is called a *region of stability*. Let

$$\bar{\mathfrak{U}} := \mathfrak{U} \cup \{ \infty \} .$$

In the study of continuous-time control systems, $\bar{\mathfrak{U}} \supset \mathbb{C}_+$.

Let the ring of proper scalar rational functions of s (with real coefficients), which have no poles in $\mathfrak{U}$ be denoted by $\mathbf{R}_{\mathfrak{U}}$. The ring $\mathbf{R}_{\mathfrak{U}}$ is a proper Euclidean domain, (which implies that it is a principal ideal domain) with degree function $\delta : \mathbf{R}_{\mathfrak{U}} \setminus 0 \to \mathbb{Z}_+$ defined by

$$\delta(f) = \text{number of } \bar{\mathfrak{U}}\text{–zeros of } f .$$

Two functions $f, g \in \mathrm{R_U}$ are coprime iff they have no common $\bar{\mathrm{U}}$–zeros. The primes in the ring $\mathrm{R_U}$ are functions of the form $\frac{s-a}{s+b}$, $\frac{1}{s+b}$, $\frac{(s-a)^2+c^2}{(s+b)^2}$ and their associates, where $a, b, c \in \mathbb{R}$, $a \in \mathrm{U}$, $-b \in \mathbb{C} \setminus \bar{\mathrm{U}}$, $c > 0$.

Suppose that $\mathrm{R_U}$ is the principal ideal domain H under consideration. By definition of J , $f \in \mathrm{J}$ implies that f is a proper rational function which has neither poles nor zeros in $\bar{\mathrm{U}}$. We choose I to be the multiplicative subset of $\mathrm{R_U}$ such that $f \in \mathrm{I}$ iff $f(\infty)$ is a nonzero constant in $\mathbb{R}$; equivalently, $\mathrm{I} \subset \mathrm{R_U}$ is the set of proper, but *not strictly proper* real rational functions which have no poles in U. The ring of fractions $\mathrm{R_U}/\mathrm{I}$ is denoted by $\mathbb{R}_p(s)$ and consists of *proper* rational functions of s with real coefficients. The field of fractions associated with the principal ideal domain $\mathrm{R_U}$ corresponds to all rational functions of s with real coefficients, denoted by $\mathbb{R}(s)$. The (Jacobson) radical of the ring $\mathbb{R}_p(s)$ is the set of strictly proper rational functions, denoted by $\mathbb{R}_{sp}(s)$.

Suppose that $p \in \mathbb{R}_p(s)$; then p can be expressed as a fraction n / d , where $n \in \mathrm{R_U}$ and $d \in \mathrm{I}$ are coprime. The $\bar{\mathrm{U}}$–zeros of p are the same as the $\bar{\mathrm{U}}$–zeros of n ; hence, $p \in \mathbb{R}_{sp}(s)$ if and only if $n / 1 = n \in \mathbb{R}_{sp}(s)$; the U–poles of p are the same as the U–zeros of d .

Let $A \in \mathfrak{m}(\mathrm{R_U})$; then A is $\mathrm{R_U}$–unimodular iff $\det A$ has no $\bar{\mathrm{U}}$–zeros, i.e., $\delta(\det A) = 0$. On the other hand, A is $\mathbb{R}_p(s)$–unimodular iff $\det A$ has no zeros at infinity.

In the case that U is chosen to be $\mathbb{C}_+$, the ring $\mathrm{R_U}$ corresponds to the set of scalar transfer functions of bounded-input-bounded-output-stable (BIBO-stable), linear, time-invariant, continuous-time, lumped systems; hence, $\mathrm{R_U}$ is called the ring of *proper stable rational functions*. Similarly, $\mathfrak{m}(\mathrm{R_U})$ is the set of matrix transfer functions of BIBO-stable systems; hence, a matrix which has elements in $\mathrm{R_U}$ is called an $\mathrm{R_U}$–stable matrix. Note that in most applications, it is desirable for the system to have poles in a *region of stability* $\mathbb{C} \setminus \mathrm{U} = \mathbb{D}$, which is more restricted than the open left-half-plane $\mathbb{C}_-$.

2.3 COPRIME FACTORIZATIONS

Suppose that $p \in \mathrm{G}$, the ring of fractions of H associated with I ; then p is equal to a fraction x / y , where $x \in \mathrm{H}$ and $y \in \mathrm{I}$. In the fraction x / y , x and y are not necessarily coprime. Suppose that $g \in \mathrm{H}$ is a greatest-common-divisor (g.c.d.) of x and y ; then there are n , $d \in \mathrm{H}$ such that $x = n\, g$ and $y = d\, g$, where $y \in \mathrm{I}$ implies that $g \in \mathrm{I}$ and $d \in \mathrm{I}$. The fraction x / y is equivalent to the fraction n / d , where $n \in \mathrm{H}$ and $d \in \mathrm{I}$ are coprime; note that since $g = u\, x + v\, y = u\, n\, g + v\, d\, g$ for some u , $v \in \mathrm{H}$, we have $u\, n + v\, d = 1$ for some u , $v \in \mathrm{H}$.

Let P be a matrix whose entries are in the ring of fractions G of the principal ideal domain H. In this section we define coprimeness in H and coprime factorizations of P over $\mathrm{m(H)}$ and display some important properties of coprime factorizations.

Definition 2.3.1. (Coprime-fraction representations)

(i) The pair (N_p , D_p) , where N_p , $D_p \in \mathrm{m(H)}$, is called *right-coprime* (**r.c.**) iff there exist U_p , $V_p \in \mathrm{m(H)}$ such that

$$V_p\, D_p + U_p\, N_p = I \; ; \tag{2.3.1}$$

(ii) the pair (N_p , D_p) is called a *right-fraction representation* (**r.f.r.**) of $P \in \mathrm{m(G)}$ iff

$$D_p \text{ is square}, \ \ \det D_p \in \mathrm{I} \ \text{ and } \ P = N_p\, D_p^{-1} \; ;$$

(iii) the pair (N_p , D_p) is called a *right-coprime-fraction representation* (**r.c.f.r.**) of $P \in \mathrm{m(G)}$ iff (N_p , D_p) is an r.f.r. of P and (N_p , D_p) is r.c.

If (N_p , D_p) is an r.c.f.r. of P then we call $N_p\, D_p^{-1}$ a *right-coprime factorization* of P.

(iv) The pair $(\tilde{D}_p, \tilde{N}_p)$, where $\tilde{D}_p, \tilde{N}_p \in \mathfrak{m}(\mathrm{H})$, is called *left-coprime* (**l.c.**) iff there exist $\tilde{U}_p, \tilde{V}_p \in \mathfrak{m}(\mathrm{H})$ such that

$$\tilde{N}_p \tilde{U}_p + \tilde{D}_p \tilde{V}_p = I \ ; \tag{2.3.2}$$

(v) the pair $(\tilde{D}_p, \tilde{N}_p)$ is called a *left-fraction representation* (**l.f.r.**) of $P \in \mathfrak{m}(\mathrm{G})$ iff

$$\tilde{D}_p \text{ is square}, \ \det\tilde{D}_p \in \mathrm{I} \quad \text{and} \quad P = \tilde{D}_p^{-1} \tilde{N}_p \ ;$$

(vi) the pair $(\tilde{D}_p, \tilde{N}_p)$ is called a *left-coprime-fraction representation* (**l.c.f.r.**) of $P \in \mathfrak{m}(\mathrm{G})$ iff $(\tilde{D}_p, \tilde{N}_p)$ is an l.f.r. of P and $(\tilde{D}_p, \tilde{N}_p)$ is l.c.

If $(\tilde{D}_p, \tilde{N}_p)$ is an l.c.f.r. of P then we call $\tilde{D}_p^{-1} \tilde{N}_p$ a *left-coprime factorization* of P.

(vii) The triple (N_{pr}, D, N_{pl}), where $N_{pr}, D, N_{pl} \in \mathfrak{m}(\mathrm{H})$, is called a *bicoprime* (**b.c.**) *triple* iff the pair (N_{pr}, D) is **r.c.** and the pair (D, N_{pl}) is **l.c.**

(viii) The quadruple (N_{pr}, D, N_{pl}, G) is called a *bicoprime-fraction representation* (**b.c.f.r.**) of $P \in \mathfrak{m}(\mathrm{G})$ iff the triple (N_{pr}, D, N_{pl}) is a bicoprime triple, D is square, $\det D \in \mathrm{I}$ and $P = N_{pr} D^{-1} N_{pl} + G$.

If (N_{pr}, D, N_{pl}, G) is a b.c.f.r. of P then we call $N_{pr} D^{-1} N_{pl} + G$ a *bicoprime factorization* of P.

□

In the factorizations $P = N_p D_p^{-1}$, $P = \tilde{D}_p^{-1} \tilde{N}_p$ and $P = N_{pr} D^{-1} N_{pl} + G$, the matrices $N_p, \tilde{N}_p, N_{pr}, N_{pl}$ are interpreted as "numerator" matrices and $D_p, \tilde{D}_p, D$ are interpreted as "denominator" matrices.

Equations (2.3.1) and (2.3.2) are called a right-Bezout identity and a left-Bezout identity, respectively.

Note that every $P \in \mathfrak{m}(\mathrm{G})$ has an r.c.f.r. (N_p, D_p), an l.c.f.r. $(\tilde{D}_p, \tilde{N}_p)$ and a b.c.f.r. (N_{pr}, D, N_{pl}, G) in H because H is a principal ring [Vid.1, Section 4.1].

Lemma 2.3.2. (Coprimeness after elementary operations)

Let $\begin{bmatrix} Y_p \\ X_p \end{bmatrix} = E \begin{bmatrix} D_p \\ N_p \end{bmatrix}$ and let $\begin{bmatrix} \tilde{X}_p & \tilde{Y}_p \end{bmatrix} = \begin{bmatrix} \tilde{N}_p & \tilde{D}_p \end{bmatrix} F$, where E , $F \in \mathrm{m}(\mathrm{H})$ are H–unimodular; then

(i) the pair (N_p , D_p) is r.c. if and only if the pair (X_p , Y_p) is r.c.;

(ii) the pair ($\tilde{D}_p$, $\tilde{N}_p$) is l.c. if and only if the pair ($\tilde{Y}_p$, $\tilde{X}_p$) is l.c.

Proof

(i) From Definition 2.3.1 (i), (N_p , D_p) is r.c. iff there exist U_p , $V_p \in \mathrm{m}(\mathrm{H})$ such that

$$\begin{bmatrix} V_p & U_p \end{bmatrix} \begin{bmatrix} D_p \\ N_p \end{bmatrix} = I = \begin{bmatrix} V_p & U_p \end{bmatrix} E^{-1} \begin{bmatrix} Y_p \\ X_p \end{bmatrix} =: \begin{bmatrix} V_y & U_x \end{bmatrix} \begin{bmatrix} Y_p \\ X_p \end{bmatrix} ;$$

equivalently, since $E^{-1} \in \mathrm{m}(\mathrm{H})$, there exist U_x, $V_y \in \mathrm{m}(\mathrm{H})$ such that $V_y Y_p + U_x X_p = I$; equivalently, the pair (X_p , Y_p) is r.c.

The proof of (ii) is entirely similar. □

Lemma 2.3.3. (Products of units)

(i) Let a , $b \in \mathrm{H}$; then $a\, b \in \mathrm{J}$ if and only if $a \in \mathrm{J}$ and $b \in \mathrm{J}$.

(ii) Let c , $d \in \mathrm{H}$; then $c\, d \in \mathrm{I}$ if and only if $c \in \mathrm{I}$ and $d \in \mathrm{I}$.

Proof

(i) If a , $b \in \mathrm{J}$, then a^{-1} , $b^{-1} \in \mathrm{H}$; since H is commutative, $1 = a^{-1} a\, b\, b^{-1} = (a\, b)(b^{-1} a^{-1}) = (b^{-1} a^{-1})(a\, b)$; hence, $b^{-1}a^{-1} \in \mathrm{H}$ is the inverse of $a\, b \in \mathrm{H}$ and $b^{-1} a^{-1} = a^{-1} b^{-1}$; therefore $a\, b \in \mathrm{J}$. To show the converse, let $a\, b =: u$; by assumption, $u^{-1} \in \mathrm{H}$. Therefore, $b \in \mathrm{H}$ has the inverse $(u^{-1} a) \in \mathrm{H}$ since $(u^{-1} a)b = 1 = b(u^{-1} a)$, and hence, $b \in \mathrm{J}$. Similarly, $a(b\, u^{-1}) = 1 = (b\, u^{-1})a$ implies that $(b\, u^{-1}) \in \mathrm{H}$ is the inverse of $a \in \mathrm{H}$ and hence, $a \in \mathrm{J}$.

(ii) Since I is a multiplicative subset of H , c , $d \in \mathrm{I}$ implies that $c\, d \in \mathrm{I}$. To show the converse, let $c\, d =: v$; then $v^{-1} \in \mathrm{G}$ since $v \in \mathrm{I}$. Now

$(v^{-1}c)d = 1 = d(v^{-1}c)$, which implies that $(v^{-1}c) \in \mathrm{G}$ is the inverse (in G) of $d \in \mathrm{H}$. Similarly, $c(dv^{-1}) = 1 = (dv^{-1})c$ implies that $(dv^{-1}) \in \mathrm{G}$ is the inverse (in G) of $c \in \mathrm{H}$; therefore, $c \in \mathrm{H}$ and $d \in \mathrm{H}$ are units in G and hence, $c \in \mathrm{I}$ and $d \in \mathrm{I}$.

Lemma 2.3.4. (Uniqueness of coprime factorizations)

Let (N_p, D_p) be an r.c.f.r. and let $(\tilde{D}_p, \tilde{N}_p)$ be an l.c.f.r. of $P \in \mathrm{m}(\mathrm{G})$; then

(i) (X_p, Y_p) is also an r.f.r. (r.c.f.r.) of P if and only if $(X_p, Y_p) = (N_p R, D_p R)$ for some G–unimodular (H–unimodular, respectively) $R \in \mathrm{m}(\mathrm{H})$;

(ii) $(\tilde{Y}_p, \tilde{X}_p)$ is also an l.f.r. (l.c.f.r.) of P if and only if $(\tilde{Y}_p, \tilde{X}_p) = (L\tilde{D}_p, L\tilde{N}_p)$ for some G–unimodular (H–unimodular, respectively) $L \in \mathrm{m}(\mathrm{H})$.

Proof

(i) *(if)* If $R \in \mathrm{m}(\mathrm{H})$ is G–unimodular, then $\det R \in \mathrm{I}$. By Definition 2.3.1 (i), $\det D_p \in \mathrm{I}$ since (N_p, D_p) is an r.c.f.r. of P; if $Y_p = D_p R$ then by Lemma 2.3.3 (ii), $\det Y_p = \det D_p \det R \in \mathrm{I}$. Since $X_p Y_p^{-1} = N_p R R^{-1} D_p^{-1} = N_p D_p^{-1} = P$, it follows that (X_p, Y_p) is an r.f.r. of P.

So far we showed that if both N_p and D_p are post-multiplied by a G–unimodular matrix R, then the resulting pair $(X_p, Y_p) = (N_p R, D_p R)$ is also an r.f.r. of P. Now if $R \in \mathrm{m}(\mathrm{H})$ is H–unimodular, i.e., if $R^{-1} \in \mathrm{m}(\mathrm{H})$, then by the (right-) Bezout identity (2.3.1) we obtain

$$R^{-1} V_p D_p R + R^{-1} U_p N_p R = R^{-1} V_p Y_p + R^{-1} U_p X_p = I \; ; \qquad (2.3.3)$$

therefore (X_p, Y_p) is also r.c. and hence, (X_p, Y_p) is an r.c.f.r. of P.

(only if) Let (X_p, Y_p) be an r.f.r. of P; then by Definition 2.3.1 (ii), $\det Y_p \in \mathrm{I}$ and hence, $Y_p^{-1} \in \mathrm{m}(\mathrm{G})$; since (N_p, D_p) is an r.c.f.r. of P, we know that $N_p D_p^{-1} = X_p Y_p^{-1}$. Now $\det D_p \in \mathrm{I}$ implies that $D_p^{-1} \in \mathrm{m}(\mathrm{G})$; post-multiplying both sides of the (right-) Bezout identity (2.3.1) by D_p^{-1}, substituting $X_p Y_p^{-1}$ for $N_p D_p^{-1} = P$ and then post-multiplying both sides by Y_p we obtain

$$V_p Y_p + U_p X_p = D_p^{-1} Y_p =: R \ , \tag{2.3.4}$$

where $R \in \mathrm{m}(\mathrm{H})$ since the left-hand side of equation (2.3.4) is a matrix with entries in H and $R^{-1} = Y_p^{-1} D_p \in \mathrm{G}$; hence $R \in \mathrm{m}(\mathrm{H})$ is G–unimodular. From equation (2.3.4), $Y_p = D_p R$ and hence, $X_p = N_p D_p^{-1} Y_p = N_p R$.

If the pair (X_p , Y_p) is also r.c., then there are matrices V_y , $U_x \in \mathrm{m}(\mathrm{H})$, such that

$$V_y Y_p + U_x X_p = I \ ; \tag{2.3.5}$$

post-multiplying both sides of (2.3.5) by Y_p^{-1} , substituting $N_p D_p^{-1}$ for $X_p Y_p^{-1} = P$ and then post-multiplying both sides by D_p we obtain

$$V_y D_p + U_x N_p = Y_p^{-1} D_p = R^{-1} \in \mathrm{m}(\mathrm{H}) \ . \tag{2.3.6}$$

Since $R \in \mathrm{m}(\mathrm{H})$ and $R^{-1} \in \mathrm{m}(\mathrm{H})$ by equations (2.3.4) and (2.3.6), we conclude that $R \in \mathrm{m}(\mathrm{H})$ is an H–unimodular matrix when (X_p , Y_p) is r.c.

The proof of (ii) is entirely similar. □

We now consider coprime factorizations of an $(\eta_o + n_o) \times (\eta_i + n_i)$ matrix $\hat{P} \in \mathrm{m}(\mathrm{G})$ partitioned as

$$\hat{P} = \begin{bmatrix} P_{11} & P_{12} \\ P_{21} & P \end{bmatrix} \in \mathrm{G}^{(\eta_o+n_o)\times(\eta_i+n_i)} \ , \quad \text{where } P \in \mathrm{G}^{n_o \times n_i} \ . \tag{2.3.7}$$

Lemma 2.3.5. (Denominator matrices in triangular form)

Let $\hat{P} \in \mathrm{m}(\mathrm{G})$ be as in equation (2.3.7); then there exist $N_{11} \in \mathrm{H}^{\eta_o \times \eta_i}$, $N_{12} \in \mathrm{H}^{\eta_o \times n_i}$, $N_{21} \in \mathrm{H}^{n_o \times \eta_i}$, $N_p \in \mathrm{H}^{n_o \times n_i}$, $D_{11} \in \mathrm{H}^{\eta_i \times \eta_i}$, $D_{21} \in \mathrm{H}^{n_i \times \eta_i}$, $D_p \in \mathrm{H}^{n_i \times n_i}$, and $\tilde{D}_{11} \in \mathrm{H}^{\eta_o \times \eta_o}$, $\tilde{D}_{12} \in \mathrm{H}^{\eta_o \times n_o}$, $\tilde{D}_p \in \mathrm{H}^{n_o \times n_o}$, $\tilde{N}_{11} \in \mathrm{H}^{\eta_o \times \eta_i}$, $\tilde{N}_{12} \in \mathrm{H}^{\eta_o \times n_i}$, $\tilde{N}_{21} \in \mathrm{H}^{n_o \times \eta_i}$, $\tilde{N}_p \in \mathrm{H}^{n_o \times n_i}$ such that

$$(N_{\hat{P}}, D_{\hat{P}}) =: \left(\begin{bmatrix} N_{11} & N_{12} \\ N_{21} & N_p \end{bmatrix}, \begin{bmatrix} D_{11} & 0 \\ D_{21} & D_p \end{bmatrix} \right) \text{ is an r.c.f.r. of } \hat{P} \tag{2.3.8}$$

and

$$(\tilde{D}_{\hat{P}}, \tilde{N}_{\hat{P}}) =: \left(\begin{bmatrix} \tilde{D}_{11} & \tilde{D}_{12} \\ 0 & \tilde{D}_p \end{bmatrix}, \begin{bmatrix} \tilde{N}_{11} & \tilde{N}_{12} \\ \tilde{N}_{21} & \tilde{N}_p \end{bmatrix} \right) \text{ is an l.c.f.r. of } \hat{P}, \tag{2.3.9}$$

where

(N_p, D_p) is an r.f.r. of P, and $(\tilde{D}_p, \tilde{N}_p)$ is an l.f.r. of P.

Comment 2.3.6.

(i) In Lemma 2.3.5 it is only claimed that (N_p, D_p) is an r.f.r. and $(\tilde{D}_p, \tilde{N}_p)$ is an l.f.r. of the 2-2 sub-block P of $\hat{P}$; these fraction representations of P are *not* necessarily *coprime*. However, $\begin{bmatrix} N_{12} \\ N_p \end{bmatrix}$ is right-coprime with D_p by equation (2.3.8) and $\tilde{D}_p$ is left-coprime with $\begin{bmatrix} \tilde{N}_{21} & \tilde{N}_p \end{bmatrix}$ by equation (2.3.9).

(ii) Let $\hat{P} = N_{\hat{P}} D_{\hat{P}}^{-1} = \tilde{D}_{\hat{P}}^{-1} \tilde{N}_{\hat{P}}$, where $(N_{\hat{P}}, D_{\hat{P}})$ is an r.c. pair as in equation (2.3.8) and $(\tilde{D}_{\hat{P}}, \tilde{N}_{\hat{P}})$ is an l.c. pair as in equation (2.3.9); then

$$\hat{P} = N_{\hat{P}} D_{\hat{P}}^{-1} = \begin{bmatrix} N_{11} & N_{12} \\ N_{21} & N_p \end{bmatrix} \begin{bmatrix} D_{11}^{-1} & 0 \\ -D_p^{-1} D_{21} D_{11}^{-1} & D_p^{-1} \end{bmatrix} \tag{2.3.10}$$

and

$$\hat{P} = \tilde{D}_{\hat{P}}^{-1} \tilde{N}_{\hat{P}} = \begin{bmatrix} \tilde{D}_{11}^{-1} & -\tilde{D}_{11}^{-1} \tilde{D}_{12} \tilde{D}_p^{-1} \\ 0 & \tilde{D}_p^{-1} \end{bmatrix} \begin{bmatrix} \tilde{N}_{11} & \tilde{N}_{12} \\ \tilde{N}_{21} & \tilde{N}_p \end{bmatrix}. \tag{2.3.11}$$

Proof of Lemma 2.3.5

Since $\hat{P} \in \mathrm{m}(\mathrm{G})$, it has an r.c.f.r. over H (call it (X , Y)) and an l.c.f.r. over H (call it $(\tilde{Y} , \tilde{X})$).

(i) By the existence of the Hermite (column-) form [Vid.1, Appendix B], there exists an H–unimodular matrix $R \in \mathrm{m}(\mathrm{H})$ such that $D_{\hat{p}} := Y R \in \mathrm{m}(\mathrm{H})$ is in the lower-(block-) triangular form given in equation (2.3.8), where we choose to denote the 2-2 entry of $D_{\hat{p}}$ by D_p ; note that $\det(Y R) = \det Y \det R = \det D_{\hat{p}} \in \mathrm{I}$. Note also that D_{11} and D_p are lower-triangular though this is not needed in the proof. Let $N_{\hat{p}} := X R \in \mathrm{m}(\mathrm{H})$, where we denote the sub-blocks in $N_{\hat{p}}$ as in equation (2.3.8), with $N_p \in \mathrm{m}(\mathrm{H})$ as the 2-2 sub-block. Since $R \in \mathrm{m}(\mathrm{H})$ is H–unimodular, by Lemma 2.3.4 (i), $(N_{\hat{p}} , D_{\hat{p}})$ is also an r.c.f.r. of $\hat{P}$.

Now equation (2.3.8) implies that $\det(Y R) = \det D_{\hat{p}} = \det D_{11} \det D_p \in \mathrm{I}$; hence by Lemma 2.3.3 (ii), $\det D_{11} \in \mathrm{I}$ and $\det D_p \in \mathrm{I}$; since $P = N_p D_p^{-1}$ by equations (2.3.7)-(2.3.8), we conclude that (N_p , D_p) is an r.f.r. of P .

(ii) Equation (2.3.9) can be justified similarly: pre-multiplying $\tilde{Y}$ by an H–unimodular $L \in \mathrm{m}(\mathrm{H})$, we obtain $\det \tilde{D}_{\hat{p}} := L \tilde{Y}$ in the upper-(block-) triangular Hermite row-form of equation (2.3.9); by Lemma 2.3.4 (ii), $(\tilde{D}_{\hat{p}} , \tilde{N}_{\hat{p}})$ is also an l.c.f.r. of $\hat{P}$. Since $\det L \det \tilde{Y} = \det \tilde{D}_{\hat{p}} = \det \tilde{D}_{11} \det \tilde{D}_p \in \mathrm{I}$, by Lemma 2.3.3 (ii), $\det \tilde{D}_p \in \mathrm{I}$; hence we conclude that $(\tilde{D}_p , \tilde{N}_p)$ is an l.f.r. of P . □

Lemma 2.3.7. (Generalized Bezout Identity)

Let (N_p , D_p) be an r.c. pair and let $(\tilde{D}_p , \tilde{N}_p)$ be an l.c. pair, and let $\tilde{N}_p D_p = \tilde{D}_p N_p$, where $N_p \in \mathrm{H}^{n_o \times n_i}$, $D_p \in \mathrm{H}^{n_i \times n_i}$, $\tilde{D}_p \in \mathrm{H}^{n_o \times n_o}$, $\tilde{N}_p \in \mathrm{H}^{n_o \times n_i}$; then there are matrices $V_p \in \mathrm{H}^{n_i \times n_i}$, $U_p \in \mathrm{H}^{n_i \times n_o}$, $\tilde{U}_p \in \mathrm{H}^{n_i \times n_o}$, $\tilde{V}_p \in \mathrm{H}^{n_o \times n_o}$ such that

$$\begin{bmatrix} V_p & U_p \\ -\tilde{N}_p & \tilde{D}_p \end{bmatrix} \begin{bmatrix} D_p & -\tilde{U}_p \\ N_p & \tilde{V}_p \end{bmatrix} = \begin{bmatrix} I_{n_i} & 0 \\ 0 & I_{n_o} \end{bmatrix} . \tag{2.3.12}$$

(Equation (2.3.12) is called a generalized Bezout identity.)

Proof

Since (N_p , D_p) is r.c. and $(\tilde{D}_p , \tilde{N}_p)$ is l.c., where $-\tilde{N}_p D_p + \tilde{D}_p N_p = 0$, there are matrices U_p, V_p, $\tilde{U}$, $\tilde{V} \in \mathrm{m}(\mathrm{H})$ such that $V_p D_p + U_p N_p = I_{n_i}$ and $\tilde{N}_p \tilde{U} + \tilde{D}_p \tilde{V} = I_{n_o}$; then

$$\begin{bmatrix} V_p & U_p \\ -\tilde{N}_p & \tilde{D}_p \end{bmatrix} \begin{bmatrix} D_p & -\tilde{U} \\ N_p & \tilde{V} \end{bmatrix} \begin{bmatrix} I_{n_i} & V_p \tilde{U} - U_p \tilde{V} \\ 0 & I_{n_o} \end{bmatrix} = \begin{bmatrix} I_{n_i} & 0 \\ 0 & I_{n_o} \end{bmatrix} .$$

Let $\tilde{U}_p := D_p (V_p \tilde{U} - U_p \tilde{V}) - \tilde{U}$ and $\tilde{V}_p := N_p (V_p \tilde{U} - U_p \tilde{V}) + \tilde{V}$; then $\tilde{U}_p$, $\tilde{V}_p \in \mathrm{m}(\mathrm{H})$ and hence, equation (2.3.12) follows. □

Corollary 2.3.8. (Generalized Bezout identities associated with bicoprime triples)

Let (N_{pr} , D , N_{pl}) be a b.c. triple, where $N_{pr} \in \mathrm{H}^{n_o \times n}$, $D \in \mathrm{H}^{n \times n}$, $N_{pl} \in \mathrm{H}^{n \times n_i}$; then we have two generalized Bezout identities:

(i) For the r.c. pair (N_{pr} , D) there are matrices $V_{pr} \in \mathrm{H}^{n \times n}$, $U_{pr} \in \mathrm{H}^{n \times n_o}$, $\tilde{X} \in \mathrm{H}^{n_o \times n}$, $\tilde{Y} \in \mathrm{H}^{n_o \times n_o}$, $\tilde{U} \in \mathrm{H}^{n \times n_o}$, $\tilde{V} \in \mathrm{H}^{n_o \times n_o}$ such that

$$\begin{bmatrix} V_{pr} & U_{pr} \\ -\tilde{X} & \tilde{Y} \end{bmatrix} \begin{bmatrix} D & -\tilde{U} \\ N_{pr} & \tilde{V} \end{bmatrix} = \begin{bmatrix} I_n & 0 \\ 0 & I_{n_o} \end{bmatrix} ; \qquad (2.3.13)$$

(ii) For the l.c. pair (D , N_{pl}) there are matrices $V_{pl} \in \mathrm{H}^{n \times n}$, $U_{pl} \in \mathrm{H}^{n_i \times n}$, $X \in \mathrm{H}^{n \times n_i}$, $Y \in \mathrm{H}^{n_i \times n_i}$, $U \in \mathrm{H}^{n_i \times n}$, $V \in \mathrm{H}^{n \times n}$ such that

$$\begin{bmatrix} D & -N_{pl} \\ U & V \end{bmatrix} \begin{bmatrix} V_{pl} & X \\ -U_{pl} & Y \end{bmatrix} = \begin{bmatrix} I_n & 0 \\ 0 & I_{n_i} \end{bmatrix} . \qquad (2.3.14)$$

Proof

(i) Since (N_{pr} , D) is an r.c. pair, there exists an H–unimodular H–stable matrix, which we call M_r , such that $\begin{bmatrix} D \\ N_{pr} \end{bmatrix}$ can be put in the Hermite form $M_r \begin{bmatrix} D \\ N_{pr} \end{bmatrix} = \begin{bmatrix} I_n \\ 0 \end{bmatrix}$. Partition M_r and label it as $\begin{bmatrix} V_{pr} & U_{pr} \\ -\tilde{X} & \tilde{Y} \end{bmatrix}$; since M_r is H–unimodular, ($\tilde{Y}$, $\tilde{X}$) is an l.c. pair; note that $\tilde{X}\, D = \tilde{Y}\, N_{pr}$. Following similar steps as in the proof of Lemma 2.3.7, there exist $\tilde{U}$, $\tilde{V}$ $\in \mathfrak{m}(\mathrm{H})$ such that equation (2.3.13) is satisfied. Note that equation (2.3.13) is of the form

$$M_r\, M_r^{-1} = I_{n+n_o} \,. \tag{2.3.15}$$

(ii) Since (D , N_{pl}) is an l.c. pair, there exists an H–unimodular H–stable matrix, which we call M_l , such that $\begin{bmatrix} D & -N_{pl} \end{bmatrix}$ can be put in the Hermite form $\begin{bmatrix} D & -N_{pl} \end{bmatrix} M_l = \begin{bmatrix} I_n & 0 \end{bmatrix}$. Equation (2.3.14) follows along similar steps as in the proof of Lemma 2.3.7 after partitioning M_l and labeling it as $\begin{bmatrix} V_{pl} & X \\ -U_{pl} & Y \end{bmatrix}$ and noting that (X , Y) is an r.c. pair, where $D\, X = N_{pl}\, Y$. Note that equation (2.3.14) is of the form

$$M_l^{-1}\, M_l = I_{n+n_i} \,. \tag{2.3.16}$$

□

Remark 2.3.9. (Doubly-coprime-fraction representations)

(i) Suppose that $Q \in \mathfrak{m}(\mathrm{H})$ is any arbitrary matrix whose entries are in H; then the generalized Bezout identity (2.3.12) implies that

$$\begin{bmatrix} V_p - Q\tilde{N}_p & U_p + Q\tilde{D}_p \\ -\tilde{N}_p & \tilde{D}_p \end{bmatrix} \begin{bmatrix} D_p & -(\tilde{U}_p + D_p Q) \\ N_p & \tilde{V}_p - N_p Q \end{bmatrix} = \begin{bmatrix} I_{n_i} & 0 \\ 0 & I_{n_o} \end{bmatrix} \tag{2.3.17}$$

equation (2.3.17) is of the form

$$M\ M^{-1} = I_{n_i + n_o} , \tag{2.3.18}$$

where $M \in \mathfrak{m}(\mathrm{H})$ is H–unimodular.

(ii) The pair $((N_p, D_p), (\tilde{D}_p, \tilde{N}_p))$ in the generalized Bezout identity (2.3.12) is called a *doubly-coprime* pair. We do not need to assume that D_p and $\tilde{D}_p$ are invertible matrices in writing equation (2.3.12). Now if (N_p, D_p) is an r.c.f.r. of P and $(\tilde{D}_p, \tilde{N}_p)$ is an l.c.f.r. of P, then the pair $((N_p, D_p), (\tilde{D}_p, \tilde{N}_p))$ in (2.3.12) is called a *doubly-coprime-fraction representation* of P and $N_p D_p^{-1} = \tilde{D}_p^{-1} \tilde{N}_p$ is called a *doubly-coprime factorization* of P; note that in this case, by Definition 2.3.1 (ii) and (v), D_p and $\tilde{D}_p$ are invertible matrices and furthermore, $D_p^{-1} \in \mathfrak{m}(\mathrm{G})$ and $\tilde{D}_p^{-1} \in \mathfrak{m}(\mathrm{G})$.

(iii) The pair $((N_{pr}, D), (\tilde{Y}, \tilde{X}))$ in equation (2.3.13) and the pair $((X, Y), (D, N_{pl}))$ in equation (2.3.14) are also doubly-coprime pairs; note that we do not need to assume that D is an invertible matrix in writing equations (2.3.13) and (2.3.14). However, if (N_{pr}, D, N_{pl}, G) is a b.c.f.r. of $P \in \mathfrak{m}(\mathrm{G})$, where $G \in \mathfrak{m}(\mathrm{H})$, then by Definition 2.3.1 (viii), D is an invertible matrix and furthermore, $D^{-1} \in \mathfrak{m}(\mathrm{G})$. □

2.4 RELATIONSHIPS BETWEEN COPRIME FACTORIZATIONS

Let (N_{pr}, D, N_{pl}, G) be a b.c.f.r. of $P \in \mathrm{m}(\mathrm{G})$. In Theorem 2.4.1 we obtain an r.c.f.r. (N_p, D_p) and an l.c.f.r. $(\tilde{D}_p, \tilde{N}_p)$ for P from (N_{pr}, D, N_{pl}, G). In Example 2.4.3 we apply Theorem 2.4.1 to the state-space representation of a matrix P that has rational function entries.

Theorem 2.4.1. (Doubly-coprime factorizations from bicoprime factorizations)

Let $P \in \mathrm{m}(\mathrm{G})$. Let (N_{pr}, D, N_{pl}, G) be a b.c.f.r. of P; hence, equations (2.3.13)-(2.3.14) hold for some $V_{pr}, U_{pr}, \tilde{X}, \tilde{Y}, \tilde{V}, \tilde{U}, V_{pl}, U_{pl}, X, Y, U, V \in \mathrm{m}(\mathrm{H})$. Under these assumptions,

$$(N_{pr} X + G Y, Y) =: (N_p, D_p) \text{ is an r.c.f.r. of } P, \tag{2.4.1}$$

$$(\tilde{Y}, \tilde{X} N_{pl} + \tilde{Y} G) =: (\tilde{D}_p, \tilde{N}_p) \text{ is an l.c.f.r. of } P. \tag{2.4.2}$$

Comment 2.4.2. (Generalized Bezout identity for the doubly-coprime pair which is obtained from a bicoprime triple)

Let (N_{pr}, D, N_{pl}) be a bicoprime triple; let $G \in \mathrm{m}(\mathrm{H})$; let $V_{pr}, U_{pr}, \tilde{X}, \tilde{Y}, \tilde{V}, \tilde{U}, V_{pl}, U_{pl}, X, Y, U, V \in \mathrm{m}(\mathrm{H})$ be as in equations (2.3.13)-(2.3.14); then $((N_{pr} X + G Y, Y), (\tilde{Y}, \tilde{X} N_{pl} + \tilde{Y} G))$ is a doubly-coprime pair. A generalized Bezout identity for this doubly-coprime pair can be obtained from equations (2.3.13)-(2.3.14) and can be verified by direct calculation:

$$\begin{bmatrix} V + U V_{pr} N_{pl} - U U_{pr} G & U U_{pr} \\ -\tilde{X} N_{pl} - \tilde{Y} G & \tilde{Y} \end{bmatrix} \begin{bmatrix} Y & -U_{pl} \tilde{U} \\ N_{pr} X + G Y & \tilde{V} + N_{pr} V_{pl} \tilde{U} - G U_{pl} \tilde{U} \end{bmatrix}$$

$$= \begin{bmatrix} I_{n_i} & 0 \\ 0 & I_{n_o} \end{bmatrix}. \tag{2.4.3}$$

Note the similarity between equations (2.3.12) and (2.4.3). Equation (2.4.3) is of the form

$$\hat{M}\ \hat{M}^{-1} = I_{n_i + n_o} \ . \qquad (2.4.4)$$

Proof of Theorem 2.4.1

By assumption, $P = N_{pr}\, D^{-1} N_{pl} + G$ and equations (2.3.13)-(2.3.14) hold. Clearly $N_{pr}\, X + G\, Y\, , Y\, , \tilde{Y}\, , \tilde{X}\, N_{pl} + \tilde{Y}\, G \in \mathrm{m(H)}$. We must show that $(N_{pr}\, X + G\, Y\, , Y\,)$ is an r.c. pair, where det$Y \in \mathrm{I}$ and that $(\tilde{Y}\, , \tilde{X}\, N_{pl} + \tilde{Y}\, G\,)$ is an l.c. pair, where det$\tilde{Y} \in \mathrm{I}$:

Equation (2.4.3) implies that $(N_{pr}\, X + G\, Y\, , Y\,)$ is an r.c. pair and $(\tilde{Y}\, , \tilde{X}\, N_{pl} + \tilde{Y}\, G\,)$ is an l.c. pair; more specifically, if $(N_{pr}\, X + G\, Y\, , Y\,) =: (N_p\, , D_p\,)$ and $(\tilde{Y}\, , \tilde{X}\, N_{pl} + \tilde{Y}\, G\,)$ $=: (\tilde{D}_p\, , \tilde{N}_p\,)$ as in equations (2.4.1)-(2.4.2), then

$$V_p\, D_p + U_p\, N_p = I_{n_i} \ , \qquad \tilde{N}_p\, \tilde{U}_p + \tilde{D}_p\, \tilde{V}_p = I_{n_o} \ , \qquad (2.4.5)$$

where

$$\begin{aligned} V_p &:= V + U\, V_{pr}\, N_{pl} - U\, U_{pr}\, G \ , \qquad U_p := U\, U_{pr} \ , \\ \tilde{U}_p &:= U_{pl}\, \tilde{U} \ , \qquad \tilde{V}_p := \tilde{V} + N_{pr}\, V_{pl}\, \tilde{U} - G\, U_{pl}\, \tilde{U} \ . \end{aligned} \qquad (2.4.6)$$

Now from equations (2.3.13)-(2.3.16), since $M_r\, M_r^{-1} = M_r^{-1}\, M_r = I_{n + n_o}$ and $M_l^{-1}\, M_r =$ $M_l\, M_l^{-1} = I_{n + n_i}$, we obtain

$$\det D = \det\left(\begin{bmatrix} D & 0 \\ 0 & I_{n_o} \end{bmatrix} M_r\, M_r^{-1} \right) = \det\left(\begin{bmatrix} I_n - \tilde{U}\tilde{X} & \tilde{U} \\ -\tilde{X} & I_{n_o} \end{bmatrix} \begin{bmatrix} I_n & 0 \\ 0 & \tilde{Y} \end{bmatrix} M_r^{-1} \right)$$

$$= \det\tilde{Y}\ \det M_r^{-1} \ , \qquad (2.4.7)$$

$$\det D = \det\left(M_l^{-1}\, M_l \begin{bmatrix} D & 0 \\ 0 & I_{n_i} \end{bmatrix} \right) = \det\left(M_l^{-1} \begin{bmatrix} I_n & 0 \\ 0 & Y \end{bmatrix} \begin{bmatrix} I_n - X\, U & X \\ -U & I_{n_i} \end{bmatrix} \right)$$

$$= \det M_l^{-1}\ \det Y \ . \qquad (2.4.8)$$

Since M_r , $M_l \in \mathrm{m}(\mathrm{H})$ are H–unimodular matrices, $\det M_r \in \mathrm{J}$ and $\det M_l^{-1} \in \mathrm{J}$; furthermore, since $\det D \in \mathrm{I}$ by assumption, equation (2.4.8) implies that

$$\det Y = \det M_l \det D \in \mathrm{I} \tag{2.4.9}$$

and equation (2.4.7) implies that

$$\det \tilde{Y} = \det M_r \det D \in \mathrm{I} . \tag{2.4.10}$$

Now by equation (2.3.14), $N_{pl}\, Y = D\, X$; hence,

$$P\, Y = (N_{pr}\, D^{-1} N_{pl} + G\,) Y = N_{pr}\, X + G\, Y . \tag{2.4.11}$$

Similarly, by equation (2.3.13), $\tilde{Y}\, N_{pr} = \tilde{X}\, D$; hence,

$$\tilde{Y}\, P = \tilde{Y}\, (N_{pr}\, D^{-1} N_{pl} + G\,) = \tilde{X}\, N_{pl} + \tilde{Y}\, G . \tag{2.4.12}$$

From equations (2.4.9)-(2.4.10), we see that $Y^{-1} \in \mathrm{m}(\mathrm{G})$ and $\tilde{Y}^{-1} \in \mathrm{m}(\mathrm{G})$; therefore, equations (2.4.11)-(2.4.12) imply that

$$P = (N_{pr}\, X + G\, Y\,) Y^{-1} = \tilde{Y}^{-1} (\tilde{X}\, N_{pl} + \tilde{Y}\, G\,) ; \tag{2.4.13}$$

therefore $(N_{pr}\, X + G\, Y , Y)$ is an r.c.f.r. of P and $(\tilde{Y} , \tilde{X}\, N_{pl} + \tilde{Y}\, G)$ is an l.c.f.r. of P .

Example 2.4.3. (Doubly-coprime factorizations from state-space representations)

Let H be $\mathrm{R_u}$ as in Section 2.2. Let $P \in \mathbb{R}_p(s)^{n_o \times n_i}$ be represented by its state-space representation $(\bar{A} , \bar{B} , \bar{C} , \bar{E})$, where $\bar{A} \in \mathbb{R}^{n \times n}$, $\bar{B} \in \mathbb{R}^{n \times n_i}$, $\bar{C} \in \mathbb{R}^{n_o \times n}$ and $\bar{E} \in \mathbb{R}^{n_o \times n_i}$. Let $-a \in \mathbb{R} \cap \mathbb{C} \backslash \bar{\mathrm{u}}$; then $P = (s + a)^{-1} \bar{C} \, [(s + a)^{-1} (s I_n - \bar{A})]^{-1} \bar{B} + \bar{E} = \bar{C} (s I_n - \bar{A}) \bar{B} + \bar{E}$. Let $(\bar{A} , \bar{B})$ be $\bar{\mathrm{u}}$–stabilizable and $(\bar{C} , \bar{A})$ be $\bar{\mathrm{u}}$–detectable; then the pair $((s + a)^{-1} \bar{C} , (s + a)^{-1} (s I_n - \bar{A}))$ is r.c. over $\mathrm{m}(\mathrm{R_u})$ and the pair $((s + a)^{-1} (s I_n - \bar{A}) , \bar{B})$ is l.c. over $\mathrm{m}(\mathrm{R_u})$; furthermore $\det [(s + a)^{-1} (s I_n - \bar{A})] \in \mathrm{I}$. Therefore,

$$(N_{pr} , D , N_{pl} , G) := ((s + a)^{-1} \bar{C} , (s + a)^{-1} (s I_n - \bar{A}) , \bar{B} , \bar{E})$$

is a b.c.f.r. of P over $\mathbf{m}(\mathrm{R_u})$. Choose $K \in \mathbb{R}^{n_i \times n}$ and $F \in \mathbb{R}^{n \times n_o}$ such that $(\bar{A} - \bar{B} K)$ and $(\bar{A} - F \bar{C})$ have all of their eigenvalues in $\mathbb{C} \setminus \bar{\mathcal{U}}$. Let

$$A_k := (s I_n - \bar{A} + \bar{B} K)^{-1} \quad \text{and} \quad A_f := (s I_n - \bar{A} + F \bar{C})^{-1} \; ; \tag{2.4.14}$$

note that $A_k , A_f \in \mathbf{m}(\mathrm{R_u}) \cap \mathbf{m}(\mathbb{R}_{sp}(s))$; since $-a \in \mathbb{C} \setminus \bar{\mathcal{U}}$, the matrices $(s+a)(s I_n - \bar{A} + \bar{B} K)^{-1} = (s+a) A_k \in \mathbf{m}(\mathrm{R_u})$ and $(s+a)(s I_n - \bar{A} + F \bar{C})^{-1} = (s+a) A_f \in \mathbf{m}(\mathrm{R_u})$ are $\mathrm{R_u}$–unimodular. For this special b.c. triple $((s+a)^{-1} \bar{C} , (s+a)^{-1} (s I - \bar{A}) , \bar{B})$, equations (2.3.13) and (2.3.14) become:

$$\begin{bmatrix} (s+a) A_f & (s+a) A_f F \\ -\bar{C} A_f & I_{n_o} - \bar{C} A_f F \end{bmatrix} \begin{bmatrix} (s+a)^{-1} (s I_n - \bar{A}) & -F \\ (s+a)^{-1} \bar{C} & I_{n_o} \end{bmatrix} = I_{n+n_o} \tag{2.4.15}$$

$$\begin{bmatrix} (s+a)^{-1} (s I_n - \bar{A}) & -\bar{B} \\ (s+a)^{-1} K & I_{n_i} \end{bmatrix} \begin{bmatrix} (s+a) A_k & (s+a) A_k \bar{B} \\ -K A_k & I_{n_i} - K A_k \bar{B} \end{bmatrix} = I_{n+n_i} . \tag{2.4.16}$$

Matching the entries of equations (2.4.15) and (2.4.16) with those of (2.3.13) and (2.3.14), respectively, we obtain a generalized Bezout identity for this special case from equation (2.4.3):

$$\begin{bmatrix} I_{n_i} + K A_f \bar{B} - K A_f F \bar{E} & K A_f F \\ -\bar{C} A_f \bar{B} - (I_{n_o} - \bar{C} A_f F) \bar{E} & I_{n_o} - \bar{C} A_f F \end{bmatrix} \begin{bmatrix} I_{n_i} - K A_k \bar{B} & -K A_k F \\ \bar{C} A_k \bar{B} + \bar{E} (I_{n_i} - K A_k \bar{B}) & I_{n_o} + \bar{C} A_k F - \bar{E} K A_k F \end{bmatrix}$$

$$= I_{n_i + n_o} . \tag{2.4.17}$$

Comparing the generalized Bezout identities (2.3.12) and (2.4.17), $(\bar{C} A_k \bar{B} + \bar{E} (I_{n_i} - K A_k \bar{B}) , (I_{n_i} - K A_k \bar{B}))$ is an r.c. pair and $((I_{n_o} - \bar{C} A_f F) , \bar{C} A_f \bar{B} + (I_{n_o} - \bar{C} A_f F) \bar{E})$ is an l.c. pair, where

$$U_p = K A_f F \quad , \quad V_p = I_{n_i} + K A_f \bar{B} - K A_f F \bar{E} \; ,$$

$$\tilde{U}_p = K A_k F \quad , \quad \tilde{V}_p = I_{n_o} + \bar{C} A_k F - \bar{E} K A_k F \ .$$

Since A_k , $A_f \in \mathfrak{m}(\mathbb{R}_{sp}(s))$, we have $\det(I_{n_i} - K A_k \bar{B}) \in \mathbf{I}$ and $\det(I_{n_o} - \bar{C} A_f F) \in \mathbf{I}$. Furthermore, since $(s+a)A_k$ is $\mathbf{R_U}$–unimodular, $\det(I_{n_i} - K A_k \bar{B}) = \det[(s+a)^{-1}(s I_n - \bar{A})]\det[(s+a)A_k] \sim \det[(s+a)^{-1}(s I_n - \bar{A})]$; similarly, since $(s+a)A_f$ is $\mathbf{R_U}$–unimodular, $\det(I_{n_o} - \bar{C} A_f F) = \det[(s+a)^{-1}(s I_n - \bar{A})]\det[(s+a)A_f] \sim \det[(s+a)^{-1}(s I_n - \bar{A})]$; therefore $\det(I_{n_i} - K A_k \bar{B}) \sim \det(I_{n_o} - \bar{C} A_f F)$. We conclude that

$$(N_p , D_p) := (\bar{C} A_k \bar{B} + \bar{E}(I_{n_i} - K A_k \bar{B}) \ , \ (I_{n_i} - K A_k \bar{B})) \quad (2.4.18)$$

is an r.c.f.r. of P over $\mathfrak{m}(\mathbf{R_U})$ and

$$(\tilde{D}_p , \tilde{N}_p) := ((I_{n_o} - \bar{C} A_f F) \ , \ \bar{C} A_f \bar{B} + (I_{n_o} - \bar{C} A_f F)\bar{E}) \quad (2.4.19)$$

is an l.c.f.r. of P over $\mathfrak{m}(\mathbf{R_U})$. □

Let (N_p , D_p) , $(\tilde{D}_p , \tilde{N}_p)$, $(N_{pr} , D , N_{pl} , G)$ be any r.c.f.r., l.c.f.r. and b.c.f.r. of $P \in \mathfrak{m}(\mathbf{G})$. By Lemma 2.3.4, any other r.c.f.r. is of the form $(N_p R , D_p R)$, where $R \in \mathfrak{m}(\mathbf{H})$ is $\mathbf{H}$–unimodular and any other l.c.f.r. is of the form $(L \tilde{D}_p , L \tilde{N}_p)$, where $L \in \mathfrak{m}(\mathbf{H})$ is $\mathbf{H}$–unimodular. By Theorem 2.4.1, any r.c.f.r. $(N_p , D_p) = ((N_{pr} X + G Y)R , Y R)$ for some $\mathbf{H}$–unimodular $R \in \mathfrak{m}(\mathbf{H})$ and any l.c.f.r. $(\tilde{D}_p , \tilde{N}_p) = (L \tilde{Y} , L(\tilde{X} N_{pl} + \tilde{Y} G))$ for some $\mathbf{H}$–unimodular $L \in \mathfrak{m}(\mathbf{H})$.

Suppose that $((N_p , D_p), (\tilde{D}_p , \tilde{N}_p))$ is a doubly-coprime pair as in equation (2.3.12) and that $(N_p , D_p) = ((N_{pr} X + G Y)R , Y R)$ for some $\mathbf{H}$–unimodular $R \in \mathfrak{m}(\mathbf{H})$, where X , $Y \in \mathfrak{m}(\mathbf{H})$ are as in equation (2.3.14). In Lemma 2.4.4 below, we show that $\det D_p$, $\det \tilde{D}_p$ and $\det D$ are associates; thus, if any one of $\det D_p$, $\det \tilde{D}_p$, $\det D$ is in $\mathbf{I}$, then the other two are also in $\mathbf{I}$. Consequently, the determinants of any r.c.f.r., any l.c.f.r. and any b.c.f.r. of $P \in \mathfrak{m}(\mathbf{G})$ are associates.

Lemma 2.4.4. (Determinants of denominator matrices of coprime factorizations)

Let $((N_p, D_p), (\tilde{D}_p, \tilde{N}_p))$ be a doubly-coprime pair as in equation (2.3.12); let $G \in \mathrm{m}(\mathrm{H})$ and let $(N_p, D_p) = ((N_{pr} X + G Y) R, Y R)$ for some H–unimodular $R \in \mathrm{m}(\mathrm{H})$, where equations (2.3.13)-(2.3.14) hold. Under these assumptions,

$$\det D_p \sim \det \tilde{D}_p \sim \det D \quad , \tag{2.4.20}$$

and for all $Q \in \mathrm{m}(\mathrm{H})$,

$$\det(\tilde{V}_p - N_p Q) \sim \det(V_p - Q \tilde{N}_p) \quad ; \tag{2.4.21}$$

furthermore,

$$\det[(V_p - Q \tilde{N}_p) D_p] = \det[\tilde{D}_p (\tilde{V}_p - N_p Q)] \ . \tag{2.4.22}$$

Proof

Since (N_p, D_p) and $(\tilde{D}_p, \tilde{N}_p)$ satisfy equation (2.3.12), equations (2.3.17)-(2.3.18) hold for all $Q \in \mathrm{m}(\mathrm{H})$; by equation (2.3.17),

$$\begin{bmatrix} D_p & 0 \\ 0 & I_{n_o} \end{bmatrix} M = \begin{bmatrix} I_{n_i} - (\tilde{U}_p + D_p Q)\tilde{N}_p & \tilde{U}_p + D_p Q \\ -\tilde{N}_p & I_{n_o} \end{bmatrix} \begin{bmatrix} I_{n_i} & 0 \\ 0 & \tilde{D}_p \end{bmatrix} . \tag{2.4.23}$$

Taking determinants of both sides of equation (2.4.23) we obtain

$$\det D_p \det M = \det \tilde{D}_p \quad . \tag{2.4.24}$$

Since $M \in \mathrm{m}(\mathrm{H})$ is H–unimodular, $\det M \in \mathrm{J}$; therefore equation (2.4.24) implies that

$$\det D_p \sim \det \tilde{D}_p \quad . \tag{2.4.25}$$

Now by Theorem 2.4.1, $(N_{pr} X + G Y, Y)$ is an r.c. pair and $(\tilde{Y}, \tilde{X} N_{pl} + \tilde{Y} G)$ is an l.c. pair since (N_{pr}, D, N_{pl}) is a b.c. triple by (2.3.13)-(2.3.14) and $G \in \mathrm{m}(\mathrm{H})$. By assumption, $(N_p, D_p) = ((N_{pr} X + G Y) R, Y R)$ for some H–unimodular $R \in \mathrm{m}(\mathrm{H})$; therefore, by equation (2.4.8), which is obtained from equations (2.3.14) and (2.3.16), we see that

$$\det D_p \;=\; \det Y \det R \;=\; \det M_l \; \det D \; \det R \quad ; \tag{2.4.26}$$

since $\det M_l \in \mathbf{J}$ and $\det R \in \mathbf{J}$, equation (2.4.26) implies that

$$\det D_p \quad \sim \quad \det D \quad . \tag{2.4.27}$$

Finally, equation (2.4.20) follows from equations (2.4.25) and (2.4.27).

Now by equations (2.3.17)-(2.3.18),

$$\begin{bmatrix} I_{n_i} & 0 \\ 0 & \tilde{V}_p - N_p Q \end{bmatrix} M = \begin{bmatrix} I_{n_i} & U_p + Q \tilde{D}_p \\ -N_p & I_{n_o} - N_p (U_p + Q \tilde{D}_p) \end{bmatrix} \begin{bmatrix} V_p - Q \tilde{N}_p & 0 \\ 0 & I_{n_o} \end{bmatrix} . \tag{2.4.28}$$

Taking determinants of both sides of equation (2.4.28) we obtain

$$\det(\tilde{V}_p \; - \; N_p \, Q \;) \det M \;\; = \;\; \det(V_p \; - \; Q \, \tilde{N}_p \;) \;\; ; \tag{2.4.29}$$

since $M \in \mathfrak{m}(\mathrm{H})$ is H–unimodular, equation (2.4.21) follows from (2.4.29). Now multiplying both sides of equation (2.4.29) by $\det D_p$ and using equation (2.4.24) we obtain

$$\det(\tilde{V}_p - N_p Q) \det M \det D_p = \det(\tilde{V}_p - N_p Q) \det \tilde{D}_p = \det(V_p - Q \tilde{N}_p) \det D_p \;\; ; \tag{2.4.30}$$

hence equation (2.4.22) follows since $\det(V_p - Q \; \tilde{N}_p) \det D_p \; = \; \det [(V_p \; - \; Q \; \tilde{N}_p) D_p]$ and $\det \tilde{D}_p \; \det(\tilde{V}_p \; - \; N_p \, Q \;) = \det [\tilde{D}_p \; (\tilde{V}_p \; - \; N_p \, Q \;)]$. □

Corollary 2.4.5. ($N_p \in \mathfrak{m}(\mathrm{G}_s)$) **implies that** $\det D_p \in \mathbf{I}$)

Let ((N_p , D_p) , ($\tilde{D}_p$, $\tilde{N}_p$)) be a doubly-coprime pair satisfying the generalized Bezout identity (2.3.12); let $N_p \in \mathfrak{m}(\mathrm{G}_s)$; Under these assumptions,

$$\det D_p \in \mathbf{I} \qquad \text{and} \qquad \det \tilde{D}_p \in \mathbf{I} \;\; ; \tag{2.4.31}$$

furthermore, for all $Q \in \mathrm{H}$,

$$\det(V_p \; - \; Q \, \tilde{N}_p \;) \in \mathbf{I} \qquad \text{and} \qquad \det(\tilde{V}_p \; - \; N_p \, Q \;) \in \mathbf{I} \; . \tag{2.4.32}$$

Proof

By assumption, (2.3.12) holds; therefore, (2.3.17) also holds for all $Q \in \mathrm{m}(\mathrm{H})$. Now $N_p \in \mathrm{m}(\mathrm{G}_s)$ implies that $(U_p + Q\tilde{D}_p)N_p \in \mathrm{m}(\mathrm{G}_s)$ for all $Q \in \mathrm{m}(\mathrm{H})$; therefore $(I_{n_i} - (U_p + Q\tilde{D}_p)N_p)^{-1} \in \mathrm{m}(\mathrm{G})$; equivalently, $\det(I_{n_i} - (U_p + Q\tilde{D}_p)N_p)$ $\in \mathrm{I}$; then by equation (2.3.17),

$$\det[(V_p - Q\tilde{N}_p)D_p] = \det[I_{n_i} - (U_p + Q\tilde{D}_p)N_p] \in \mathrm{I}. \qquad (2.4.33)$$

By Lemma 2.3.3 (ii), equation (2.4.33) holds if and only if $\det D_p \in \mathrm{I}$ and $\det(\tilde{V}_p - Q\tilde{N}_p) \in \mathrm{I}$; since $\det M \in \mathrm{J}$, equations (2.4.31)-(2.4.32) follow from equations (2.4.24) and (2.4.29). □

Lemma 2.4.6. (Denominator matrices of H–stable matrices)

Let (N_{pr}, D, N_{pl}, G) be a b.c.f.r. of $P \in \mathrm{m}(\mathrm{G})$; then $P \in \mathrm{m}(\mathrm{H})$ if and only if $D^{-1} \in \mathrm{m}(\mathrm{H})$; equivalently, $\det D \in \mathrm{J}$.

Proof

If $D^{-1} \in \mathrm{m}(\mathrm{H})$ then $P = N_{pr} D^{-1} N_{pl} + G \in \mathrm{m}(\mathrm{H})$ since $N_{pr}, N_{pl}, G \in \mathrm{m}(\mathrm{H})$. To show the converse, let $N_{pr} D^{-1} N_{pl} + G \in \mathrm{m}(\mathrm{H})$; then $N_{pr} D^{-1} N_{pl} = P - G \in \mathrm{m}(\mathrm{H})$. By equation (2.3.14), $N_{pr} D^{-1} N_{pl} U_{pl} = N_{pr} D^{-1}(I_n - D V_{pl}) = N_{pr} D^{-1} - N_{pr} V_{pl} \in \mathrm{m}(\mathrm{H})$ and equivalently, $N_{pr} D^{-1}$. Furthermore, by equation (2.3.13), $U_{pr} N_{pr} D^{-1} = (I_n - V_{pr} D) D^{-1} = D^{-1} - V_{pr} \in \mathrm{m}(\mathrm{H})$ and equivalently, $D^{-1} \in \mathrm{m}(\mathrm{H})$. □

Comment 2.4.7.

(i) Let (N_{pr}, D, N_{pl}) be a b.c. triple over $\mathrm{m}(\mathrm{H})$ and let $G \in \mathrm{m}(\mathrm{H})$; then following similar steps as in the proof of Lemma 2.4.6, we can easily show that $N_{pr} D^{-1} N_{pl} + G \in \mathrm{m}(\mathrm{G})$ if and only if $D^{-1} \in \mathrm{m}(\mathrm{G})$; but since $D \in \mathrm{m}(\mathrm{H})$, $D^{-1} \in \mathrm{m}(\mathrm{G})$ if and only if $\det D \in \mathrm{I}$.

(ii) Let (N_p, D_p) be an r.c.f.r. and $(\tilde{D}_p, \tilde{N}_p)$ be an l.c.f.r. of $P \in \mathrm{m}(\mathrm{G})$; then $(N_p, D_p, I, 0)$ and $(I, \tilde{D}_p, \tilde{N}_p, 0)$ are bicoprime-fraction representations of P and hence, by Lemma 2.4.6, $P \in \mathfrak{m}(\mathrm{H})$ if and only if $D_p^{-1} \in \mathfrak{m}(\mathrm{H})$ and equivalently, $\tilde{D}_p^{-1} \in \mathfrak{m}(\mathrm{H})$.

(iii) If $P \in \mathfrak{m}(\mathrm{H})$ then Lemma 2.4.6 implies that (P, I_{n_i}) is an r.c.f.r. and (I_{n_o}, P) is an l.c.f.r. of P; by Lemma 2.3.4, any other r.c.f.r. of P is of the form (PR, R), where $R \in \mathfrak{m}(\mathrm{H})$ is H–unimodular and any other l.c.f.r. of P is of the form (L, LP), where $L \in \mathfrak{m}(\mathrm{H})$ is H–unimodular.

(iv) Let (N_{pr}, D, N_{pl}, G) be a b.c.f.r. of $P \in \mathfrak{m}(\mathrm{G})$; then $\det D \in \mathrm{I}$ is a *characteristic determinant* of P [Vid.1, Section 4.3]. It follows from Lemma 2.4.4 that $\det D_p$ and $\det \tilde{D}_p$ are also characteristic determinants of P. By Lemma 2.4.6, $P \in \mathfrak{m}(\mathrm{H})$ if and only if D is H–unimodular and equivalently, the characteristic determinant of P is in the group of units J of H.

(v) Let H be the ring $\mathrm{R}_{\mathcal{U}}$ as in Section 2.2. Let $P \in \mathfrak{m}(\mathbb{R}_p(s))$; let $N_p D_p^{-1} = \tilde{D}_p^{-1} \tilde{N}_p = N_{pr} D^{-1} N_{pl} + G$ be r.c., l.c., and b.c. factorizations of P. Let the set of $\bar{\mathcal{U}}$–zeros of $\det D$ be denoted by

$$\mathcal{Z}[\det D] := \{ s_o \in \bar{\mathcal{U}} \mid \det D(s_o) = 0 \} ; \tag{2.4.34}$$

note that $\det D(\infty) \neq 0$ since $\det D \in \mathrm{I}$. Lemma 2.4.4 implies that

$$\mathcal{Z}[\det D_p] = \mathcal{Z}[\det \tilde{D}_p] = \mathcal{Z}[\det D] . \tag{2.4.35}$$

An element $d \in \bar{\mathcal{U}}$ is a $\bar{\mathcal{U}}$–pole of $P \in \mathfrak{m}(\mathbb{R}_p(s))$ iff d is an $\bar{\mathcal{U}}$–zero of a characteristic determinant of P; equivalently, $d \in \bar{\mathcal{U}}$ is an $\bar{\mathcal{U}}$–pole of P iff $d \in \mathcal{Z}[\det D] = \mathcal{Z}[\det D_p] = \mathcal{Z}[\det \tilde{D}_p]$. Note that P has no poles at infinity since $P \in \mathfrak{m}(\mathbb{R}_p(s))$.

The McMillan degree of $d \in \bar{\mathcal{U}}$ as a pole of P is, by definition, equal to its multiplicity as a zero of a characteristic determinant of P. □

2.5 ALL SOLUTIONS OF THE MATRIX EQUATIONS

$$\tilde{X}\, A = B \;, \qquad \tilde{A}\, X = \tilde{B}$$

In this section we consider all solutions for $\tilde{X}$ and X over $\mathrm{m}(\mathrm{H})$ of the matrix equations $\tilde{X}\, A = B$ and $\tilde{A}\, X = \tilde{B}$, where $\tilde{X} = \begin{bmatrix} \tilde{D}_c & \tilde{N}_c \end{bmatrix}$, $X = \begin{bmatrix} -N_c \\ D_c \end{bmatrix}$, $A = \begin{bmatrix} D_p \\ N_p \end{bmatrix}$, $\tilde{A} = \begin{bmatrix} -\tilde{N}_p & \tilde{D}_p \end{bmatrix}$ (see (2.5.3) and (2.5.4) below).

Lemma 2.5.1. (Parametrization of all solutions)

Let $((N_p\,, D_p\,), (\tilde{D}_p\,, \tilde{N}_p\,))$ be a doubly-coprime pair satisfying the generalized Bezout identity (2.3.12). Consider the equations

$$\tilde{D}_c\, D_p + \tilde{N}_c\, N_p = B \;, \tag{2.5.1}$$

and

$$\tilde{N}_p\, N_c + \tilde{D}_p\, D_c = \tilde{B} \;, \tag{2.5.2}$$

where $B \in \mathrm{H}^{n_i \times n_i}$ and $\tilde{B} \in \mathrm{H}^{n_o \times n_o}$. Under these assumptions,

(i) $(\tilde{D}_c\,, \tilde{N}_c\,)$ is a solution of equation (2.5.1) over $\mathrm{m}(\mathrm{H})$ if and only if

$$\begin{bmatrix} \tilde{D}_c & \tilde{N}_c \end{bmatrix} = \begin{bmatrix} B & Q \end{bmatrix} \begin{bmatrix} V_p & U_p \\ -\tilde{N}_p & \tilde{D}_p \end{bmatrix} \tag{2.5.3}$$

for some $Q \in \mathrm{m}(\mathrm{H})$.

(ii) $(N_c, D_c\,)$ is a solution of equation (2.5.2) over $\mathrm{m}(\mathrm{H})$ if and only if

$$\begin{bmatrix} -N_c \\ D_c \end{bmatrix} = \begin{bmatrix} D_p & -\tilde{U}_p \\ N_p & \tilde{V}_p \end{bmatrix} \begin{bmatrix} -Q \\ \tilde{B} \end{bmatrix} \tag{2.5.4}$$

for some $Q \in \mathrm{m}(\mathrm{H})$. □

Equation (2.5.3) is a parametrization of all solutions of the pair $(\tilde{D}_c, \tilde{N}_c)$ in (2.5.1) over $\mathfrak{m}(\mathrm{H})$; similarly equation (2.5.4) is a parametrization of all solutions of the pair (N_c, D_c) in (2.5.2) over $\mathfrak{m}(\mathrm{H})$.

Proof

(i) *(if)* Suppose that $(\tilde{D}_c, \tilde{N}_c)$ is as in equation (2.5.3); then by equation (2.3.12),

$$\tilde{D}_c D_p + \tilde{N}_c N_p = \begin{bmatrix} \tilde{D}_c & \tilde{N}_c \end{bmatrix} \begin{bmatrix} D_p \\ N_p \end{bmatrix} = \begin{bmatrix} B & Q \end{bmatrix} \begin{bmatrix} V_p & U_p \\ -\tilde{N}_p & \tilde{D}_p \end{bmatrix} \begin{bmatrix} D_p \\ N_p \end{bmatrix} =$$

$\begin{bmatrix} B & Q \end{bmatrix} \begin{bmatrix} I_{n_i} \\ 0 \end{bmatrix} = B$ and hence, equation (2.5.1) is satisfied.

(only if) By assumption, $(\tilde{D}_c, \tilde{N}_c)$ satisfies equation (2.5.1). Let $Q :=$ $-\tilde{D}_c \tilde{U}_p + \tilde{N}_c \tilde{V}_p \in \mathfrak{m}(\mathrm{H})$; then

$$\begin{bmatrix} \tilde{D}_c & \tilde{N}_c \end{bmatrix} \begin{bmatrix} D_p & -\tilde{U}_p \\ N_p & \tilde{V}_p \end{bmatrix} = \begin{bmatrix} B & Q \end{bmatrix} . \tag{2.5.5}$$

Post-multiplying both sides of equation (2.5.5) by the H–unimodular matrix $\begin{bmatrix} V_p & U_p \\ -\tilde{N}_p & \tilde{D}_p \end{bmatrix}$ and using equation (2.3.12), we obtain the solution given by equation (2.5.3).

The proof of part (ii) is entirely similar. □

Remark 2.5.2.

(i) Suppose that $B = I_{n_i}$ and $\tilde{B} = I_{n_o}$ in the matrix equations $\tilde{X} A = B$ and $\tilde{A} X = \tilde{B}$; then $\tilde{X}$ is the left-inverse of A over $\mathfrak{m}(\mathrm{H})$ and X is the right-inverse of $\tilde{A}$ over $\mathfrak{m}(\mathrm{H})$. In Lemma 2.5.1, if $B = I_{n_i}$, then (2.5.1) is a left-Bezout identity for the l.c. pair $(\tilde{D}_c, \tilde{N}_c)$ and if $\tilde{B} = I_{n_o}$, then (2.5.2) is a right-Bezout identity for the r.c. pair (N_c, D_c); in this case, if in addition $\tilde{D}_c N_c = \tilde{N}_c D_c$, then (2.5.1)-(2.5.2) imply that $((N_c, D_c), (\tilde{D}_c, \tilde{N}_c))$ is a doubly-coprime pair, where the associated generalized Bezout identity is:

$$\begin{bmatrix} \tilde{D}_c & \tilde{N}_c \\ -\tilde{N}_p & \tilde{D}_p \end{bmatrix} \begin{bmatrix} D_p & -N_c \\ N_p & D_c \end{bmatrix} = \begin{bmatrix} I_{n_i} & 0 \\ 0 & I_{n_o} \end{bmatrix} . \tag{2.5.6}$$

Comparing the generalized Bezout identities (2.3.17) and (2.5.6), from Lemma 2.5.1, all solutions of (2.5.6) over $\mathfrak{m}(\mathrm{H})$ are given by

$$(\tilde{D}_c, \tilde{N}_c) = ((V_p - Q\tilde{N}_p), (U_p + Q\tilde{D}_p)) , \tag{2.5.7}$$

$$(N_c, D_c) = ((\tilde{U}_p + D_p Q), (\tilde{V}_p - N_p Q)) , \tag{2.5.8}$$

where $Q \in \mathfrak{m}(\mathrm{H})$.

The matrix $Q \in \mathfrak{m}(\mathrm{H})$ in equations (2.5.7)-(2.5.8) is called a *(matrix-) parameter* in the sense that all solutions of (2.5.6) for $((N_c, D_c), (\tilde{D}_c, \tilde{N}_c))$ are parametrized by the matrix Q. Note that if $\tilde{D}_c N_c = \tilde{N}_c D_c$ as in equation (2.5.6), then the (matrix-) parameter $Q \in \mathfrak{m}(\mathrm{H})$ in (2.5.7) is the same as the (matrix-) parameter $Q \in \mathfrak{m}(\mathrm{H})$ in (2.5.8).

(ii) Suppose that $P \in \mathfrak{m}(\mathrm{G}_s)$ and that (N_p, D_p) is an r.c.f.r., $(\tilde{D}_p, \tilde{N}_p)$ is an l.c.f.r. of P. Let the generalized Bezout identity (2.3.12) hold; then $N_p = P D_p \in \mathfrak{m}(\mathrm{G}_s)$ and $\tilde{N}_p = \tilde{D}_p P \in \mathfrak{m}(\mathrm{G}_s)$. By Corollary 2.4.5, $(V_p - Q\tilde{N}_p)^{-1} \in \mathfrak{m}(\mathrm{G})$ and $(\tilde{V}_p - N_p Q)^{-1} \in \mathfrak{m}(\mathrm{G})$ for all $Q \in \mathfrak{m}(\mathrm{H})$.

With $P \in \mathfrak{m}(\mathrm{G}_s)$, suppose that the pair $((N_c, D_c), (\tilde{D}_c, \tilde{N}_c))$ satisfies the generalized Bezout identity (2.5.6); then the solutions in (2.5.7)-(2.5.8) have the property that

$$\det(V_p - Q\tilde{N}_p) \in \mathrm{I} \text{ and } \det(\tilde{V}_p - N_p Q) \in \mathrm{I} , \tag{2.5.9}$$

for all $Q \in \mathfrak{m}(\mathrm{H})$. Let $C := \tilde{D}_c^{-1}\tilde{N}_c = N_c D_c^{-1}$; then for all $Q \in \mathfrak{m}(\mathrm{H})$, $C \in \mathfrak{m}(\mathrm{G})$, where by equations (2.5.7)-(2.5.8),

$$\begin{aligned} C &= \tilde{D}_c^{-1}\tilde{N}_c = (V_p - Q\tilde{N}_p)^{-1}(U_p + Q\tilde{D}_p) \\ C &= N_c D_c^{-1} = (\tilde{U}_p + D_p Q)(\tilde{V}_p - N_p Q)^{-1} \end{aligned} \tag{2.5.10}$$

are l.c. and r.c. factorizations of $C \in \mathfrak{m}(\mathrm{G})$.

(iii) Now suppose that $P \in \mathfrak{m}(\mathbf{G})$ but not in $\mathfrak{m}(\mathbf{G}_s)$; then N_p and $\tilde{N}_p$ are not in $\mathfrak{m}(\mathbf{G}_s)$ either; consequently, $\det(V_p - Q\tilde{N}_p)$ and $\det(\tilde{V}_p - N_p Q)$ are not necessarily in $\mathbf{I}$ for all $Q \in \mathfrak{m}(\mathbf{H})$. In this case, $\tilde{D}_c = (V_p - Q\tilde{N}_p)$ and $D_c = (\tilde{V}_p - N_p Q)$ are valid denominator matrices for $C \in \mathfrak{m}(\mathbf{G})$ as in equation (2.5.10) only for those $Q \in \mathfrak{m}(\mathbf{H})$ such that condition (2.5.9) is satisfied. One (conservative) way to ensure that condition (2.5.9) will be satisfied is to choose $Q \in \mathfrak{m}(\mathbf{H})$ such that

$$\tilde{N}_c = (U_p + Q\tilde{D}_p) \in \mathfrak{m}(\mathbf{G}_s)\,; \tag{2.5.11}$$

in this case, $\det\tilde{D}_c \det D_p = \det(I_{n_i} - \tilde{N}_c N_p) \in \mathbf{I}$, which implies that $\det\tilde{D}_c = \det(V_p - Q\tilde{N}_p) \in \mathbf{I}$ and hence, condition (2.5.9) is satisfied. Note that choosing $Q \in \mathfrak{m}(\mathbf{H})$ such that (2.5.11) is satisfied guarantees $\tilde{D}_c^{-1} \in \mathfrak{m}(\mathbf{G})$ and hence, it follows that $N_c = \tilde{D}_c^{-1}\tilde{N}_c D_c$ is also in $\mathfrak{m}(\mathbf{G}_s)$.

Note that those $Q \in \mathfrak{m}(\mathbf{H})$ that satisfy (2.5.11) actually parametrize all solutions of C in equation (2.5.10) which are in $\mathfrak{m}(\mathbf{G}_s)$ since $C = \tilde{D}_c^{-1}\tilde{N}_c = N_c D_c^{-1} \in \mathfrak{m}(\mathbf{G}_s)$ if and only if $\tilde{N}_c \in \mathfrak{m}(\mathbf{G}_s)$ and equivalently, $N_c \in \mathfrak{m}(\mathbf{G}_s)$.

Choosing a matrix $Q \in \mathfrak{m}(\mathbf{H})$ that satisfies (2.5.11) can be reduced to a simple (scalar) sufficient condition as follows: Choose $q \in \mathbf{H}$ such that

$$1 + q\,(\det\tilde{D}_p) \in \mathbf{G}_s\ , \tag{2.5.12}$$

and take

$$Q^o := q\,(\det\tilde{D}_p)\,U_p\,\tilde{D}_p^{-1}\ ; \tag{2.5.13}$$

note that $(\det\tilde{D}_p)\tilde{D}_p^{-1} \in \mathbf{H}$ and hence, $Q^o \in \mathfrak{m}(\mathbf{R_U})$; Q^o satisfies condition (2.5.11) since $(1 + q\,(\det\tilde{D}_p))\,U_p \in \mathfrak{m}(\mathbf{G}_s)$.

(iv) Suppose that $\mathbf{H}$ is the ring $\mathbf{R_U}$ as in Section 2.2. If $P \in \mathfrak{m}(\mathbb{R}_{sp}(s))$, then condition (2.5.9) is satisfied for all $Q \in \mathfrak{m}(\mathbf{R_U})$. If $P \in \mathfrak{m}(\mathbb{R}_p(s))$ but not in $\mathfrak{m}(\mathbb{R}_{sp}(s))$, then condition (2.5.11), which guarantees (2.5.9), is satisfied if and only if $\tilde{N}_c = (U_p + Q\tilde{D}_p)$

$\in \mathbf{m}(\mathrm{H}) \cap \mathbf{m}(\mathbb{R}_{sp}(s))$; equivalently, $Q \in \mathbf{m}(\mathrm{R_u})$ is such that

$$Q(\infty) = -U_p(\infty)\tilde{D}_p^{-1}(\infty) , \tag{2.5.14}$$

where $\det\tilde{D}_p(\infty) \neq 0$ since $\det\tilde{D}_p \in \mathrm{I}$ by Definition 2.3.1 (ii). Choosing $Q \in \mathbf{m}(\mathrm{R_u})$ as in equation (2.5.14) is a *sufficient* condition for $(V_p - Q\tilde{N}_p)^{-1} \in \mathbf{m}(\mathbb{R}_p(s))$ and equivalently, $(\tilde{V}_p - N_p Q)^{-1} \in \mathbf{m}(\mathbb{R}_p(s))$. It is important to note that the matrices $C \in \mathbf{m}(\mathbb{R}_p(s))$ that satisfy equation (2.5.10) are in $\mathbf{m}(\mathbb{R}_{sp}(s))$ if and only if $Q \in \mathbf{m}(\mathrm{R_u})$ is chosen as in (2.5.14).

(v) Suppose that $P \in \mathbf{m}(\mathrm{H})$; then following the discussion in Comment 2.4.7 (iii), an r.c.f.r. can be chosen as (P, I_{n_i}) and an l.c.f.r. can be chosen as (I_{n_o}, P); hence in the generalized Bezout identity (2.3.12), we can choose $V_p = I_{n_i}$, $\tilde{V}_p = I_{n_o}$, $U_p = \tilde{U}_p = 0$. In this case the solutions in (2.5.7)-(2.5.8) can be replaced by

$$(\tilde{D}_c, \tilde{N}_c) = ((I_{n_i} - QP), Q) , \tag{2.5.15}$$

$$(N_c, D_c) = (Q, (I_{n_o} - PQ)) , \tag{2.5.16}$$

where $Q \in \mathbf{m}(\mathrm{H})$.

If $P \in \mathbf{m}(\mathrm{H}) \cap \mathbf{m}(\mathrm{G}_s)$, then

$$\det(I_{n_i} - QP) = \det(I_{n_o} - PQ) \in \mathrm{I} \tag{2.5.17}$$

for all $Q \in \mathbf{m}(\mathrm{H})$; therefore,

$$C := (I_{n_i} - QP)^{-1}Q = Q(I_{n_o} - PQ)^{-1} \in \mathbf{m}(\mathrm{G}) , \tag{2.5.18}$$

for all $Q \in \mathbf{m}(\mathrm{H})$; furthermore, $C \in \mathbf{m}(\mathrm{G}_s)$ if and only if $Q \in \mathbf{m}(\mathrm{H}) \cap \mathbf{m}(\mathrm{G}_s)$.

If $P \in \mathbf{m}(\mathrm{H})$ but not in $\mathbf{m}(\mathrm{G}_s)$, then choosing $Q \in \mathbf{m}(\mathrm{H}) \cap \mathbf{m}(\mathrm{G}_s)$ is sufficient to satisfy (2.5.17).

(vi) In Lemma 2.5.1, suppose that we started with an l.c. pair $(\tilde{D}_c, \tilde{N}_c)$ and an r.c. pair (N_c, D_c) satisfy the following generalized Bezout identity:

$$\begin{bmatrix} V_c & U_c \\ -\tilde{N}_c & \tilde{D}_c \end{bmatrix} \begin{bmatrix} D_c & -\tilde{U}_c \\ N_c & \tilde{V}_c \end{bmatrix} = \begin{bmatrix} I_{n_o} & 0 \\ 0 & I_{n_i} \end{bmatrix}. \tag{2.5.19}$$

In this case, equation (2.5.1) is of the form $\tilde{A}X = B$ and equation (2.5.2) is of the form $\tilde{X}A = \tilde{B}$, where $X = \begin{bmatrix} -N_p \\ D_p \end{bmatrix}$, $\tilde{X} = \begin{bmatrix} \tilde{D}_p & \tilde{N}_p \end{bmatrix}$, $\tilde{A} = \begin{bmatrix} -\tilde{N}_c & \tilde{D}_c \end{bmatrix}$, $A = \begin{bmatrix} D_c \\ N_c \end{bmatrix}$. Under these assumptions, (N_p, D_p) is a solution of equation (2.5.1) over $\mathrm{m}(\mathrm{H})$ if and only if

$$\begin{bmatrix} -N_p \\ D_p \end{bmatrix} = \begin{bmatrix} D_c & -\tilde{U}_c \\ N_c & \tilde{V}_c \end{bmatrix} \begin{bmatrix} -Q_p \\ B \end{bmatrix}, \tag{2.5.20}$$

for some $Q_p \in \mathrm{m}(\mathrm{H})$; similarly, $(\tilde{D}_p, \tilde{N}_p)$ is a solution of equation (2.5.2) over $\mathrm{m}(\mathrm{H})$ if and only if

$$\begin{bmatrix} \tilde{D}_p & \tilde{N}_p \end{bmatrix} = \begin{bmatrix} \tilde{B} & Q_p \end{bmatrix} \begin{bmatrix} V_c & U_c \\ -\tilde{N}_c & \tilde{D}_c \end{bmatrix}, \tag{2.5.21}$$

for some $Q_p \in \mathrm{m}(\mathrm{H})$. □

2.6 RANK CONDITIONS FOR COPRIMENESS

In this section, the principal ideal domain $\mathbf{H}$ under consideration is the ring $\mathbf{R_U}$ of proper stable rational functions as in Section 2.2.

Lemma 2.6.1. (Rank test for right- or left-coprimeness)

(i) Let $N_p \in \mathbf{R_U}^{n_o \times n_i}$, $D_p \in \mathbf{R_U}^{n_i \times n_i}$; then (N_p , D_p) is r.c. if and only if

$$rank \begin{bmatrix} D_p(s) \\ N_p(s) \end{bmatrix} = n_i \ , \quad \text{for all } s \in \bar{\mathbf{U}} \ . \tag{2.6.1}$$

(ii) Let $\tilde{D}_p \in \mathbf{R_U}^{n_o \times n_o}$, $\tilde{N}_p \in \mathbf{R_U}^{n_o \times n_i}$; then ($\tilde{D}_p$, $\tilde{N}_p$) is l.c. if and only if

$$rank \begin{bmatrix} \tilde{D}_p(s) & \tilde{N}_p(s) \end{bmatrix} = n_o \ , \quad \text{for all } s \in \bar{\mathbf{U}} \ . \tag{2.6.2}$$

□

Note that the rank tests for right-coprimeness and left-coprimeness in (2.6.1) and (2.6.2), respectively, need to be performed only at the $\bar{\mathbf{U}}$–zeros of detD_p (equivalently, at the $\bar{\mathbf{U}}$–zeros of det$\tilde{D}_p$), since these rank conditions hold automatically for all other $s \in \bar{\mathbf{U}}$.

Proof

(i) (N_p , D_p) is an r.c. pair if and only if there is an $\mathbf{R_U}$–unimodular matrix E (labeled as $\begin{bmatrix} V_p & U_p \\ -\tilde{N}_p & \tilde{D}_p \end{bmatrix}$) such that

$$\begin{bmatrix} V_p(s) & U_p(s) \\ -\tilde{N}_p(s) & \tilde{D}_p(s) \end{bmatrix} \begin{bmatrix} D_p(s) \\ N_p(s) \end{bmatrix} = \begin{bmatrix} I_{n_i} \\ 0 \end{bmatrix} ; \tag{2.6.3}$$

since the matrix E has rank $n_i + n_o$ for all $s \in \bar{\mathbf{U}}$, equation (2.6.3) holds if and only if the rank condition (2.6.1) holds.

(ii) Similar to part (i): the pair $(\tilde{D}_p, \tilde{N}_p)$ is l.c. if and only if there is an $\mathbf{R_U}$–unimodular matrix F (labeled as $\begin{bmatrix} D_p & -\tilde{U}_p \\ N_p & \tilde{V}_p \end{bmatrix}$) such that

$$\begin{bmatrix} -\tilde{N}_p(s) & \tilde{D}_p(s) \end{bmatrix} \begin{bmatrix} D_p(s) & -\tilde{U}_p(s) \\ N_p(s) & \tilde{V}_p(s) \end{bmatrix} = \begin{bmatrix} 0 & I_{n_o} \end{bmatrix} ; \qquad (2.6.4)$$

since the matrix F has rank $n_i + n_o$ for all $s \in \bar{\mathcal{U}}$, equation (2.6.4) holds if and only if the rank condition (2.6.2) holds. □

Let

$$\max_{K \in \mathcal{K}} \; rank\, M(K) \qquad (2.6.5)$$

denote the maximum rank that the matrix $M(K)$ achieves as K varies over the set $\mathcal{K}$.

Lemma 2.6.2

Let $A \in \mathbb{C}^{\eta \times \gamma}$, $B \in \mathbb{C}^{\rho \times \gamma}$, $\tilde{A} \in \mathbb{C}^{\tilde{\rho} \times \tilde{\gamma}}$, $\tilde{B} \in \mathbb{C}^{\tilde{\rho} \times \tilde{\eta}}$ be complex constant matrices.

(i) If for all $K \in \mathbb{R}^{\rho \times \eta}$,

$$rank \begin{bmatrix} B + KA \end{bmatrix} < \min \{ \rho, \gamma \} , \qquad (2.6.6)$$

then

$$rank \begin{bmatrix} B \\ A \end{bmatrix} = \max_{K \in \mathfrak{m}(\mathbb{R})} rank \begin{bmatrix} B + KA \end{bmatrix} . \qquad (2.6.7)$$

(ii) If for all $\tilde{K} \in \mathbb{R}^{\tilde{\gamma} \times \tilde{\eta}}$,

$$rank \begin{bmatrix} \tilde{B} + \tilde{A}\tilde{K} \end{bmatrix} < \min \{ \tilde{\rho}, \tilde{\eta} \} , \qquad (2.6.8)$$

then

$$rank \begin{bmatrix} \tilde{B} & \tilde{A} \end{bmatrix} = \max_{\tilde{K} \in \mathfrak{m}(\mathbb{R})} rank \begin{bmatrix} \tilde{B} + \tilde{A}\tilde{K} \end{bmatrix} . \qquad (2.6.9)$$

Proof

We only prove part (i); the proof of (ii) is entirely similar.

(i) Let $\hat{K}$ be a $\rho \times \eta$ real matrix that maximizes $rank \begin{bmatrix} B + K A \end{bmatrix}$; let $r :=$ $rank \begin{bmatrix} B + \hat{K} A \end{bmatrix}$; by equation (2.6.6), this rank, $r < \min \{ \rho , \gamma \}$. Therefore, there is a nonsingular complex matrix $R \in \mathbb{C}^{\gamma \times \gamma}$, which corresponds to elementary column operations on complex matrices, and there is a nonsingular real matrix $L \in \mathbb{R}^{\rho \times \rho}$, which corresponds to row permutations, such that

$$L \begin{bmatrix} B + \hat{K} A \end{bmatrix} R = \begin{bmatrix} I_r & 0 \\ G & 0 \end{bmatrix}, \tag{2.6.10}$$

where G is a $(\rho - r) \times r$ complex matrix and the zero in the bottom right is $(\rho - r) \times (\gamma - r)$, with $r < \min \{ \rho , \gamma \}$.

Let $A R =: \begin{bmatrix} \bar{A} & \hat{A} \end{bmatrix}$, where $\bar{A} \in \mathbb{C}^{\eta \times r}$ and $\hat{A} \in \mathbb{C}^{\eta \times (\gamma - r)}$; then $\begin{bmatrix} L (B + \hat{K} A) R \\ A R \end{bmatrix} = \begin{bmatrix} I_r & 0 \\ G & 0 \\ \bar{A} & \hat{A} \end{bmatrix}$. Now since $\hat{K}$ is a maximizer of $rank \begin{bmatrix} B + K A \end{bmatrix}$, for all $\hat{K}_2 \in \mathbb{R}^{(\rho - r) \times \eta}$,

$$rank \left(L \left[B + \left(\hat{K} + L^{-1} \begin{bmatrix} 0 \\ \hat{K}_2 \end{bmatrix} \right) A \right] R \right) = rank \left(L (B + \hat{K} A) R + \begin{bmatrix} 0 \\ \hat{K}_2 \end{bmatrix} A R \right)$$

$$= rank \begin{bmatrix} I_r & 0 \\ G + \hat{K}_2 \bar{A} & \hat{K}_2 \hat{A} \end{bmatrix} \leq r . \tag{2.6.11}$$

By equation (2.6.11), $rank\ \hat{K}_2 \hat{A} = 0$ for all $\hat{K}_2$ and hence, $\hat{A}$ is the $\eta \times (\gamma - r)$ zero matrix.

Therefore,

$$rank \begin{bmatrix} B \\ A \end{bmatrix} = rank \left(\begin{bmatrix} L & 0 \\ 0 & I \end{bmatrix} \begin{bmatrix} B + \hat{K} A \\ A \end{bmatrix} R \right) = rank \begin{bmatrix} I_r & 0 \\ G & 0 \\ \bar{A} & 0 \end{bmatrix} = r$$

$$= rank \begin{bmatrix} B + \hat{K} A \end{bmatrix} = \max_{K \in \mathfrak{m}(\mathbb{R})} rank \begin{bmatrix} B + K A \end{bmatrix} .$$

□

Corollary 2.6.3

Let (N_p, D_p) be an r.c.f.r., $(\tilde{D}_p, \tilde{N}_p)$ be an l.c.f.r., (N_{pr}, D, N_{pl}) be a b.c.r.f. of $P \in \mathfrak{m}(\mathbb{R}_p(s))$, where N_p, D_p, $\tilde{D}_p$, $\tilde{N}_p$, N_{pr}, D, $N_{pl} \in \mathfrak{m}(\mathrm{R_U})$. Under these assumptions, for each $s_o \in \bar{\mathcal{U}}$,

(i) there exists a real constant matrix $K \in \mathbb{R}^{n_i \times n_o}$ such that

$$rank \begin{bmatrix} D_p(s_o) + K N_p(s_o) \end{bmatrix} = n_i ; \tag{2.6.12}$$

(ii) there exists a real constant matrix $\tilde{K} \in \mathbb{R}^{n_i \times n_o}$ such that

$$rank \begin{bmatrix} \tilde{D}_p(s_o) + \tilde{N}_p(s_o) \tilde{K} \end{bmatrix} = n_o . \tag{2.6.13}$$

Proof

We only prove part (i); the proof of (ii) is similar.

(i) Suppose, for a contradiction, that there is an $s_o \in \bar{\mathcal{U}}$ such that, for all $K \in \mathbb{R}^{n_i \times n_o}$,

$$rank \begin{bmatrix} D_p(s_o) + K N_p(s_o) \end{bmatrix} < n_i . \tag{2.6.14}$$

Note that $D_p(s_o) \in \mathbb{C}^{n_i \times n_i}$ and $N_p(s_o) \in \mathbb{C}^{n_o \times n_i}$. By Lemma 2.6.2, equation (2.6.14) implies that $rank \begin{bmatrix} D_p(s_o) \\ N_p(s_o) \end{bmatrix} < n_i$; but by Lemma 2.6.1, this contradicts the fact that the pair (N_p, D_p) is right-coprime. □

Chapter 3

FULL-FEEDBACK CONTROL SYSTEMS

3.1 INTRODUCTION

This chapter studies linear, time-invariant (lti), multiinput-multioutput (MIMO) control systems with full-feedback compensators: in particular, the classical unity-feedback system $\mathbf{S}(P, C)$ (see Figure 3.1) and the more general system configuration $\Sigma(\hat{P}, \hat{C})$ (see Figure 3.9) are considered.

In the unity-feedback system $\mathbf{S}(P, C)$, the plant has only one (vector-)input e and one (vector)-output y; this output is used in feedback to the compensator; hence, the plant model considers only additive inputs or disturbances (say u, u'), which all affect the plant through its actuators. More generally, however, there may be inputs (for example, disturbances, initial conditions, noise, manual commands) which are applied directly to the plant without going through the actuators; hence, the map from the directly applied inputs to the plant output may be different from the map from the additive inputs to the plant output. Furthermore, the regulated output variable of the plant (for example, tracking error, actuator states) may not be accessible (e.g. some plant states), may not be directly measured (e.g. tracking error) or may be different from the measured output (e.g. ideal sensor outputs); the measured output is utilized by the compensator. The configuration $\Sigma(\hat{P}, \hat{C})$ represents a feedback connection which takes such cases into account.

The unity-feedback configuration $\mathbf{S}(P, C)$ is studied in Section 3.2. Compensator design using the configuration $\mathbf{S}(P, C)$ is called one-degree-of-freedom design (or one-parameter design) due to the single free parameter matrix Q that parametrizes all

H–stabilizing compensators. The class of all achievable input-output (I/O) maps for $\mathbf{S}(P, C)$ is obtained by using the class of all stabilizing compensators; this class is given in equation (3.2.58); all closed-loop I/O maps in the H–stabilized $\mathbf{S}(P, C)$ are affine maps in the (matrix-) parameter Q. The problem of diagonalizing the I/O map $H_{yu'}$ in the configuration $\mathbf{S}(P, C)$ is discussed for H–stable plants in Section 3.2.15.

The important theorems in Section 3.2 are Theorem 3.2.7 (H–stability conditions for $\mathbf{S}(P, C)$ in terms of coprime factorizations of the plant and the compensator), Theorem 3.2.11 (parametrization of all compensators that H–stabilize the plant) and Theorem 3.2.16 (compensators that achieve decoupling for H–stable plants).

The system configuration $\Sigma(\hat{P}, \hat{C})$ represents the most general interconnection of two systems, a plant $\hat{P}$ and a compensator $\hat{C}$. This general system configuration is studied in Section 3.3; the plant and the compensator each have two (vector-)inputs and two (vector-)outputs. The measured output y of $\hat{P}$ is used in feedback, but the output z is the actual output of the plant (the output in the performance specifications); the output signals z and y are not the same. The input v is considered as a disturbance, noise or an external command applied directly to the plant. The compensator output y', which is utilized by the plant in feedback, can be considered as the ideal actuator inputs; the output z' of $\hat{C}$ can be used for performance monitoring or fault diagnosis. The input v' of $\hat{C}$ is considered as the independent control input; for example, commands or initial conditions. The signals u and u', which appear at the interconnection of $\hat{P}$ and $\hat{C}$ model possible additive disturbances, noise, interference and loading.

In the configuration $\Sigma(\hat{P}, \hat{C})$, intuitively only those plants which have "instabilities that the feedback-loop can remove" can be considered for H–stabilization; these plants are called Σ–admissible. The restriction on the class of H–stabilizable $\hat{P}$ is due to the feedback being applied only through the second input e and the second output y. The 2-2 block of $\hat{C}$ is essentially in a feedback configuration like $\mathbf{S}(P, C)$ of Section 3.2; hence the set of all C that H–stabilizes the feedback-loop is the same as for $\mathbf{S}(P, C)$.

The class of all $\hat{C}$ that H–stabilizes $\hat{P}$ is parametrized by *four* H–stable (matrix-) parameters and hence, compensator design using the configuration $\Sigma(\hat{P}, \hat{C})$ is called four-degrees-of-freedom design (or four-parameter design). The configuration $\Sigma(\hat{P}, \hat{C})$ can obviously be reduced to two-parameter design by taking $C_{11} = 0$ and $C_{12} = 0$; but $\Sigma(\hat{P}, \hat{C})$ is clearly much more advantageous and general than two-degrees-of-freedom design with a two-input one-output compensator [see, for example, Vid. 1]. The class of all achievable maps for $\Sigma(\hat{P}, \hat{C})$, given by equation (3.3.70), involves the four compensator (matrix-) parameters Q_{11}, Q_{12}, Q_{21}, Q; *each* closed-loop I/O map achieved by the H–stabilized $\Sigma(\hat{P}, \hat{C})$ depends on one and *only one* of these four (matrix-) parameters. Clearly, several *independent* performance specifications may be imposed on the closed-loop performance of $\Sigma(\hat{P}, \hat{C})$. For example, decoupling the I/O map $H_{zv'} = N_{12} Q_{21}$ is independent of the I/O maps that are affine functions in Q. On the other hand, in the unity-feedback configuration $S(P, C)$, diagonalizing the map $H_{yu'} : u' \mapsto y$ would depend on the choice for Q such that $N_p \, (U_p + Q \tilde{D}_p)$ is diagonal, and hence, diagonalizing the map $H_{yu'}$ in $S(P, C)$ may not be possible for certain plants.

The problem of diagonalizing the closed-loop I/O map $H_{zv'}$ from the external-input v' to the actual-output z for the plant $\hat{P}$ is solved in Section 3.3.15; the I/O map $H_{zv'}$ in the configuration $\Sigma(\hat{P}, \hat{C})$ can always be diagonalized while preserving closed-loop stability.

The achievable I/O maps of a two (vector-) input two (vector-) output plant $\hat{P}$ (described by its state-space representation) are calculated in Example 3.3.17.

The important theorems in Section 3.3 are Theorem 3.3.5 (H–stability conditions for $\Sigma(\hat{P}, \hat{C})$ in terms of coprime factorizations of $\hat{P}$ and $\hat{C}$), Theorem 3.3.9 (conditions for Σ–admissibility of $\hat{P}$) Theorem 3.3.10 (parametrization of all compensators that H–stabilize $\hat{P}$) and Theorem 3.3.15 (class of all achievable diagonal $H_{zv'}$); the compensators that diagonalize the I/O map $H_{zv'}$ from the independent control input v' to the actual output z are specified by equation (3.3.80).

3.2 THE STANDARD UNITY-FEEDBACK SYSTEM

In this section we consider the linear, time-invariant unity-feedback system $\mathbf{S}(P, C)$ shown in Figure 3.1, where $P : e \mapsto y$ represents the plant and $C : e' \mapsto y'$ represents the compensator. The externally applied inputs are denoted by $\bar{u} := \begin{bmatrix} u \\ u' \end{bmatrix}$, the inputs to the plant and the compensator are denoted by $\bar{e} := \begin{bmatrix} e \\ e' \end{bmatrix}$, the plant and the compensator outputs are denoted by $\bar{y} := \begin{bmatrix} y \\ y' \end{bmatrix}$; the closed-loop input-output (I/O) map of $\mathbf{S}(P, C)$ is denoted by $H_{\bar{y}\bar{u}} : \bar{u} \mapsto \bar{y}$.

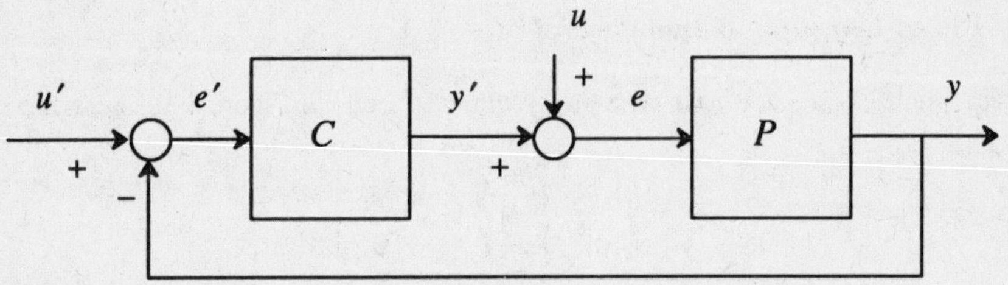

Figure 3.1. The unity-feedback system $\mathbf{S}(P, C)$.

3.2.1. Assumptions on $\mathbf{S}(P, C)$

(i) The plant $P \in \mathbf{G}^{n_o \times n_i}$.

(ii) The compensator $C \in \mathbf{G}^{n_i \times n_o}$.

(iii) The system $\mathbf{S}(P, C)$ is well-posed; equivalently, the closed-loop input-output map

$$H_{\bar{y}\bar{u}} : \begin{bmatrix} u \\ u' \end{bmatrix} \mapsto \begin{bmatrix} y \\ y' \end{bmatrix} \text{ is in } \mathbf{m}(\mathbf{G}). \qquad \square$$

Note that whenever P satisfies Assumption 3.2.1 (i), it has an r.c.f.r., denoted by (N_p, D_p), an l.c.f.r., denoted by $(\tilde{D}_p, \tilde{N}_p)$ and a b.c.f.r., denoted by (N_{pr}, D, N_{pl}, G), where $N_p \in \mathrm{H}^{n_o \times n_i}$, $D_p \in \mathrm{H}^{n_i \times n_i}$, $\tilde{D}_p \in \mathrm{H}^{n_o \times n_o}$, $\tilde{N}_p \in \mathrm{H}^{n_o \times n_i}$, $N_{pr} \in \mathrm{H}^{n_o \times n}$, $D \in \mathrm{H}^{n \times n}$, $N_{pl} \in \mathrm{H}^{n \times n_i}$. Hence, the generalized Bezout identities (2.3.12), (2.3.13), (2.3.14) are satisfied for some V_p, U_p, $\tilde{V}_p$, $\tilde{U}_p$, V_{pr}, U_{pr}, $\tilde{X}$, $\tilde{Y}$, $\tilde{V}$, $\tilde{U}$, V_{pl}, U_{pl}, X, Y, U, $V \in \mathrm{m}(\mathrm{H})$.

Similarly, whenever C satisfies Assumption 3.2.1 (ii), it has an l.c.f.r., denoted by $(\tilde{D}_c, \tilde{N}_c)$ and an r.c.f.r., denoted by (N_c, D_c), where $\tilde{D}_c \in \mathrm{H}^{n_i \times n_i}$, $\tilde{N}_c \in \mathrm{H}^{n_i \times n_o}$, $N_c \in \mathrm{H}^{n_i \times n_o}$, $D_c \in \mathrm{H}^{n_o \times n_o}$.

If $\mathbf{S}(P, C)$ is a lumped, continuous-time, linear, time-invariant system, then the principal ideal domain H under consideration is the ring of proper stable rational functions $\mathrm{R_u}$ as in Section 2.2; in that case we assume that P and $C \in \mathrm{m}(\mathbb{R}_p(s))$.

3.2.2. Closed-loop input-output maps of $\mathbf{S}(P, C)$

Let Assumptions 3.2.1 hold; then the system $\mathbf{S}(P, C)$ in Figure 3.1 is described by:

$$\bar{e} = \bar{u} + \begin{bmatrix} 0 & I_{n_i} \\ -I_{n_o} & 0 \end{bmatrix} \bar{y} \; ; \tag{3.2.1}$$

$$\bar{y} = \begin{bmatrix} P & 0 \\ 0 & C \end{bmatrix} \bar{e} \; . \tag{3.2.2}$$

Substituting for $\bar{e}$ from equation (3.2.1) into (3.2.2), we obtain

$$\left(\begin{bmatrix} I_{n_o} & 0 \\ 0 & I_{n_i} \end{bmatrix} - \begin{bmatrix} P & 0 \\ 0 & C \end{bmatrix} \begin{bmatrix} 0 & I_{n_i} \\ -I_{n_o} & 0 \end{bmatrix} \right) \bar{y} = \begin{bmatrix} P & 0 \\ 0 & C \end{bmatrix} \bar{u} \; . \tag{3.2.3}$$

Writing $\left(\begin{bmatrix} I_{n_o} & 0 \\ 0 & I_{n_i} \end{bmatrix} - \begin{bmatrix} P & 0 \\ 0 & C \end{bmatrix} \begin{bmatrix} 0 & I_{n_i} \\ -I_{n_o} & 0 \end{bmatrix} \right)^{-1} \begin{bmatrix} P & 0 \\ 0 & C \end{bmatrix}$ as

$$\begin{bmatrix} I_{n_o} & -P \\ C & I_{n_i} \end{bmatrix}^{-1} \begin{bmatrix} 0 & -I_{n_i} \\ I_{n_o} & 0 \end{bmatrix}^{-1} - \begin{bmatrix} 0 & I_{n_o} \\ -I_{n_i} & 0 \end{bmatrix} ,$$

it is easy to see that the I/O map $H_{\bar{y}\bar{u}} : \bar{u} \mapsto \bar{y}$ is in $\mathfrak{m}(\mathbf{G})$ if and only if

$$\begin{bmatrix} I_{n_o} & -P \\ C & I_{n_i} \end{bmatrix}^{-1} \in \mathfrak{m}(\mathbf{G}) \; ; \tag{3.2.4}$$

equivalently, $H_{\bar{y}\bar{u}} \in \mathfrak{m}(\mathbf{G})$ if and only if $(I_{n_i} + C\,P\,)^{-1} \in \mathfrak{m}(\mathbf{G})$ if and only if $(I_{n_o} + P\,C\,)^{-1} \in \mathfrak{m}(\mathbf{G})$. Note that $(I_{n_i} + CP\,)^{-1} \in \mathfrak{m}(\mathbf{G})$ if and only if $\det(I_{n_i} + CP\,) = \det(I_{n_o} + PC\,)$ is a unit in $\mathbf{G}$. In the case that $\mathbf{S}(P, C\,)$ is a lumped, continuous-time, linear, time-invariant system where $P\,, C \in \mathfrak{m}(\mathbb{R}_{sp}(s))$, condition (3.2.4) is equivalent to

$$\det(I_{n_i} + C(\infty)P(\infty)) = \det(I_{n_o} + P(\infty)C(\infty)) \neq 0\,. \tag{3.2.5}$$

Now since Assumption 3.2.1 (iii) holds, condition (3.2.4) is satisfied. Therefore the closed-loop I/O map $H_{\bar{y}\bar{u}} : \begin{bmatrix} u \\ u' \end{bmatrix} \mapsto \begin{bmatrix} y \\ y' \end{bmatrix}$ is in $\mathfrak{m}(\mathbf{G})$, where $H_{\bar{y}\bar{u}}$ is given in terms of $(I_{n_i} + C\,P\,)^{-1}$ in equation (3.2.6) and in terms of $(I_{n_o} + P\,C\,)^{-1}$ in equation (3.2.7) below:

$$H_{\bar{y}\bar{u}} = \begin{bmatrix} P\,(I_{n_i} + C\,P\,)^{-1} & P\,(I_{n_i} + C\,P\,)^{-1}C \\ (I_{n_i} + C\,P\,)^{-1} - I_{n_i} & (I_{n_i} + C\,P\,)^{-1}C \end{bmatrix} ; \tag{3.2.6}$$

$$H_{\bar{y}\bar{u}} = \begin{bmatrix} (I_{n_o} + P\,C\,)^{-1}P & (I_{n_o} + P\,C\,)^{-1}P\,C \\ -C\,(I_{n_o} + P\,C\,)^{-1}P & C\,(I_{n_o} + P\,C\,)^{-1} \end{bmatrix} . \tag{3.2.7}$$

The equivalent expressions (3.2.6) and (3.2.7) for $H_{\bar{y}\bar{u}}$ are obtained from (3.2.3) and using the following well-known matrix identities:

$$P\,(I_{n_i} + C\,P\,)^{-1} = (I_{n_o} + P\,C\,)^{-1}P \;\;,\;\; (I_{n_i} + C\,P\,)^{-1}C = C\,(I_{n_o} + P\,C\,)^{-1} \,, \tag{3.2.8}$$

$$I_{n_i} - (I_{n_i} + C\,P\,)^{-1}C\,P = (I_{n_i} + C\,P\,)^{-1} = I_{n_i} - C\,(I_{n_o} + P\,C\,)^{-1}P\,. \tag{3.2.9}$$

Definition 3.2.3. (H–stability of S(P , C))

The system S(P , C) is said to be H–*stable* iff $H_{\overline{yu}} \in$ m(H) . □

If S(P , C) is a lumped, continuous-time, linear, time-invariant system, then the principal ideal domain under consideration is $\mathrm{R_U}$ as in Section 2.2; therefore we say S(P , C) is $\mathrm{R_U}$–stable instead if H–stable.

Note that by Definition 3.2.3, the well-posedness of S(P , C) is a necessary condition for its H–stability. Each of the *four* transfer matrices in $H_{\overline{yu}}$ must be in m(H) for the closed-loop system to be H–stable; in the case that $P \in$ m(H) as in Lemma 3.2.4 or in the case that $C \in$ m(H) as in Lemma 3.2.5 below, checking the H–stability of S(P , C) is reduced to checking only one of the four matrices in $H_{\overline{yu}}$:

Lemma 3.2.4. (Closed-loop stability when the plant is H–stable)

Let Assumptions 3.2.1 hold and let $P \in$ m(H) ; under these assumptions, S(P , C) is H–stable if and only if

$$H_{y'u'} := (I_{n_i} + C\,P\,)^{-1} C \;\in\; \mathrm{m(H)}\ . \tag{3.2.10}$$

Proof

(only if) By Definition 3.2.1, if S(P , C) is H–stable, then $H_{\overline{yu}} \in$ m(H) ; therefore $H_{y'u'} \in$ m(H).

(if) By assumption, $H_{y'u'} \in$ m(H) and $P \in$ m(H) ; using the matrix identities (3.2.8)-(3.2.9) in the expression (3.2.6) for $H_{\overline{yu}}$ we obtain $H_{yu} := P\,(I_{n_i} + C\,P)^{-1} = P\,[I_{n_i} - (I_{n_i} + C\,P)^{-1} C\,] = P\,[I_{n_i} - H_{y'u'} P\,] \in$ m(H) ; $H_{yu'} := P\,(I_{n_i} + C\,P\,)^{-1} C = P\,H_{y'u'} \in$ m(H) ; $H_{y'u} := (I_{n_i} + C\,P\,)^{-1} - I_{n_i} = -(I_{n_i} + C\,P\,)^{-1} C\,P = -H_{y'u'} P \in$ m(H). Therefore,

$$H_{\overline{yu}} = \begin{bmatrix} H_{yu} & H_{yu'} \\ H_{y'u} & H_{y'u'} \end{bmatrix} = \begin{bmatrix} P\,(I_{n_i} - H_{yu'} P\,) & P\,H_{y'u'} \\ -H_{y'u'} P & H_{y'u'} \end{bmatrix} \in \mathrm{m(H)} \tag{3.2.11}$$

and hence, S(P , C) is H–stable. □

Lemma 3.2.5. (Closed-loop stability when the compensator is H–stable)

Let Assumptions 3.2.1 hold and let $C \in \mathrm{m}(\mathrm{H})$; under these assumptions, $\mathrm{S}(P, C)$ is H–stable if and only if

$$H_{yu} := P(I_{n_i} + C P)^{-1} \in \mathrm{m}(\mathrm{H}) . \tag{3.2.12}$$

Proof

(only if) By Definition 3.2.1, if $\mathrm{S}(P, C)$ is H–stable, then $H_{\overline{yu}} \in \mathrm{m}(\mathrm{H})$; therefore $H_{yu} \in \mathrm{m}(\mathrm{H})$.

(if) By assumption, $H_{yu} \in \mathrm{m}(\mathrm{H})$ and $C \in \mathrm{m}(\mathrm{H})$; using the matrix identities (3.2.8)-(3.2.9) in the expression (3.2.7) for $H_{\overline{yu}}$ we obtain

$$H_{\overline{yu}} = \begin{bmatrix} H_{yu} & H_{yu} C \\ -C H_{yu} & (I_{n_i} - C H_{yu}) C \end{bmatrix} \in \mathrm{m}(\mathrm{H}) \tag{3.2.13}$$

and hence, $\mathrm{S}(P, C)$ is H–stable. □

If Assumptions 3.2.1 hold and if both the plant and the compensator are H–stable, then $\mathrm{S}(P, C)$ is H–stable if and only if $(I_{n_i} + C P)^{-1} \in \mathrm{m}(\mathrm{H})$, equivalently, $(I_{n_o} + P C)^{-1} \in \mathrm{m}(\mathrm{H})$; i.e., when when $P \in \mathrm{m}(\mathrm{H})$ and $C \in \mathrm{m}(\mathrm{H})$, $\mathrm{S}(P, C)$ is H–stable if and only if $\det(I_{n_i} + C P) = \det(I_{n_o} + P C) \in \mathrm{J}$.

3.2.6. Analysis (Descriptions of $\mathrm{S}(P, C)$ using coprime factorizations)

We now analyze the unity-feedback system $\mathrm{S}(P, C)$ using coprime factorizations over $\mathrm{m}(\mathrm{H})$ of the plant and the compensator transfer matrices; this analysis leads us to the characterization of closed-loop H–stability in terms of coprime factorizations and the parametrization of all H–stabilizing compensators such that the closed-loop system is H–stable.

Assumptions 3.2.1 hold throughout this analysis.

(i) Analysis of S(P, C) with $P = N_p D_p^{-1}$ and $C = \tilde{D}_c^{-1} \tilde{N}_c$

Let (N_p, D_p) be any r.c.f.r. of $P \in \mathfrak{m}(\mathrm{G})$ and let $(\tilde{D}_c, \tilde{N}_c)$ be any l.c.f.r. of $C \in \mathfrak{m}(\mathrm{G})$. The system S(P, C) in Figure 3.1 can be redrawn as in Figure 3.2 below, where $P = N_p D_p^{-1}$ and $C = \tilde{D}_c^{-1} \tilde{N}_c$; note that $D_p \xi_p = e$, $y = N_p \xi_p$, where ξ_p denotes the *pseudo-state* of P.

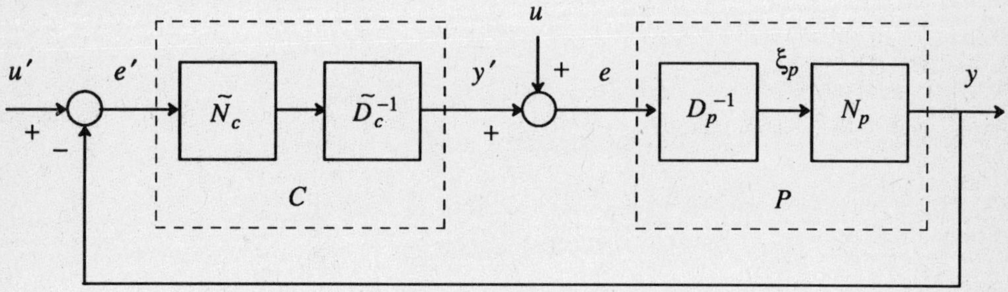

Figure 3.2. S(P, C) with $P = N_p D_p^{-1}$ and $C = \tilde{D}_c^{-1} \tilde{N}_c$.

The system S(P, C) is then described by equations (3.2.14)-(3.2.15):

$$\left[\tilde{D}_c D_p + \tilde{N}_c N_p \right] \xi_p = \left[\tilde{D}_c \quad \tilde{N}_c \right] \begin{bmatrix} u \\ u' \end{bmatrix} , \tag{3.2.14}$$

$$\begin{bmatrix} N_p \\ D_p \end{bmatrix} \xi_p = \begin{bmatrix} y \\ y' \end{bmatrix} - \begin{bmatrix} 0 & 0 \\ -I_{n_i} & 0 \end{bmatrix} \begin{bmatrix} u \\ u' \end{bmatrix} . \tag{3.2.15}$$

Equations (3.2.14)-(3.2.15) are of the form

$$D_{H1} \xi_p = N_{HL1} \bar{u}$$

$$N_{HR1} \xi_p = \bar{y} - G_{H1} \bar{u} .$$

By Lemma 2.3.2, performing elementary row operations over $\mathfrak{m}(\mathrm{H})$ on the matrix $\begin{bmatrix} D_{H1} \\ N_{HR1} \end{bmatrix}$ and elementary column operations over $\mathfrak{m}(\mathrm{H})$ on the matrix $\left[N_{HL1} \quad D_{H1} \right]$, we conclude that $(N_{HR1}, D_{H1}, N_{HL1})$ is a b.c. triple. Since $D_{H1}, G_{H1} \in \mathfrak{m}(\mathrm{H})$, it follows from Comment 2.4.7 (i) that

$$H_{\overline{yu}} = N_{HR1} D_{H1}^{-1} N_{HL1} + G_{H1} \in \mathrm{m}(\mathrm{G}) \tag{3.2.16}$$

(equivalently, the system $\mathbf{S}(P, C)$ is well-posed) if and only if $\det D_{H1} \in \mathrm{I}$. Since Assumption 3.2.1 (iii) holds, condition (3.2.16) is satisfied and hence, $\det D_{H1} \in \mathrm{I}$. Consequently, $(N_{HR1}, D_{H1}, N_{HL1}, G_{H1})$ is a b.c.f.r. of $H_{\overline{yu}}$ and hence, $\det D_{H1}$ is a characteristic determinant of $H_{\overline{yu}}$.

(ii) Analysis of $\mathbf{S}(P, C)$ with $P = \tilde{D}_p^{-1} \tilde{N}_p$ and $C = N_c D_c^{-1}$

Let $(\tilde{D}_p, \tilde{N}_p)$ be any l.c.f.r. of $P \in \mathrm{m}(\mathrm{G})$ and let (N_c, D_c) be any r.c.f.r. of $C \in \mathrm{m}(\mathrm{G})$. The system $\mathbf{S}(P, C)$ in Figure 3.1 can be redrawn as in Figure 3.3 below, where $P = \tilde{D}_p^{-1} \tilde{N}_p$ and $C = N_c D_c^{-1}$; note that $D_c \xi_c = e'$, $y' = N_c \xi_c$, where ξ_c denotes the *pseudo-state* of C.

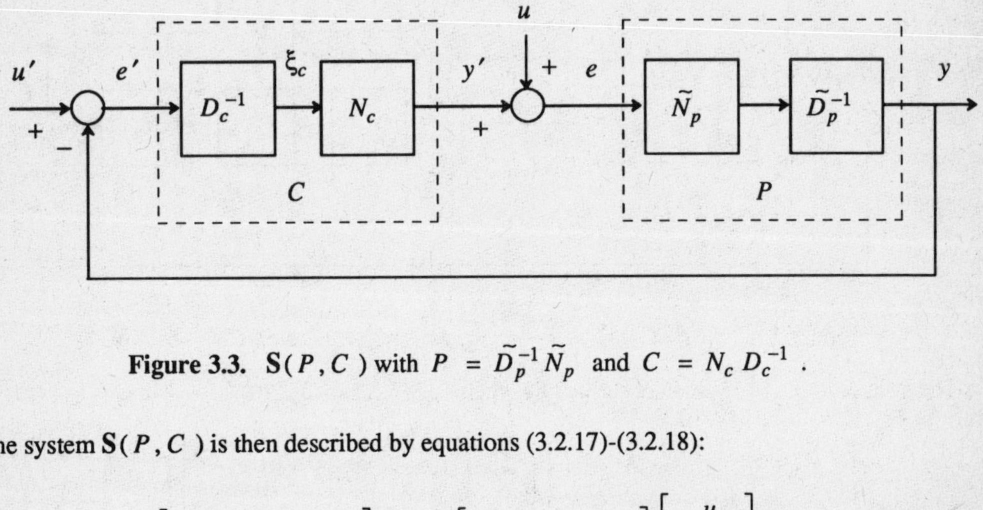

Figure 3.3. $\mathbf{S}(P, C)$ with $P = \tilde{D}_p^{-1} \tilde{N}_p$ and $C = N_c D_c^{-1}$.

The system $\mathbf{S}(P, C)$ is then described by equations (3.2.17)-(3.2.18):

$$\begin{bmatrix} \tilde{D}_p D_c + \tilde{N}_p N_c \end{bmatrix} \xi_c = \begin{bmatrix} -\tilde{N}_p & \tilde{D}_p \end{bmatrix} \begin{bmatrix} u \\ u' \end{bmatrix}, \tag{3.2.17}$$

$$\begin{bmatrix} -D_c \\ N_c \end{bmatrix} \xi_c = \begin{bmatrix} y \\ y' \end{bmatrix} - \begin{bmatrix} 0 & I_{n_o} \\ 0 & 0 \end{bmatrix} \begin{bmatrix} u \\ u' \end{bmatrix}. \tag{3.2.18}$$

Equations (3.2.17)-(3.2.18) are of the form

$$D_{H2}\,\xi_c = N_{HL2}\,\bar{u}$$
$$N_{HR2}\,\xi_c = \bar{y} - G_{H2}\,\bar{u} \ .$$

As in Analysis 3.2.6 (i) above, it can be easily verified that $(N_{HR2}, D_{H2}, N_{HL2})$ is a b.c. triple and that $H_{\bar{y}\bar{u}} = N_{HR2} D_{H2}^{-1} N_{HL2} + G_{H2} \in \mathrm{m}(\mathrm{G})$ if and only if $\det D_{H2} \in \mathrm{I}$. Again by Assumption 3.2.1 (iii), $H_{\bar{y}\bar{u}} \in \mathrm{m}(\mathrm{G})$ and hence, $\det D_{H2} \in \mathrm{I}$. Consequently, $(N_{HR2}, D_{H2}, N_{HL2}, G_{H2})$ is a b.c.f.r. of $H_{\bar{y}\bar{u}}$ and hence, $\det D_{H2}$ is a characteristic determinant of $H_{\bar{y}\bar{u}}$.

(iii) Analysis of $\mathbf{S}(P, C)$ with $P = N_{pr} D^{-1} N_{pl} + G$ and $C = \tilde{D}_c^{-1} \tilde{N}_c$

Let (N_{pr}, D, N_{pl}, G) be any b.c.f.r. of $P \in \mathrm{m}(\mathrm{G})$ and let $(\tilde{D}_c, \tilde{N}_c)$ be any l.c.f.r. of $C \in \mathrm{m}(\mathrm{G})$. The system $\mathbf{S}(P, C)$ in Figure 3.1 can be redrawn as in Figure 3.4 below, where $P = N_{pr} D^{-1} N_{pl} + G$ and $C = \tilde{D}_c^{-1} \tilde{N}_c$; note that $D\,\xi_x = N_{pl}\,e$, $y = N_{pr}\,\xi_x + G\,e$, where ξ_x denotes the *pseudo-state* of P.

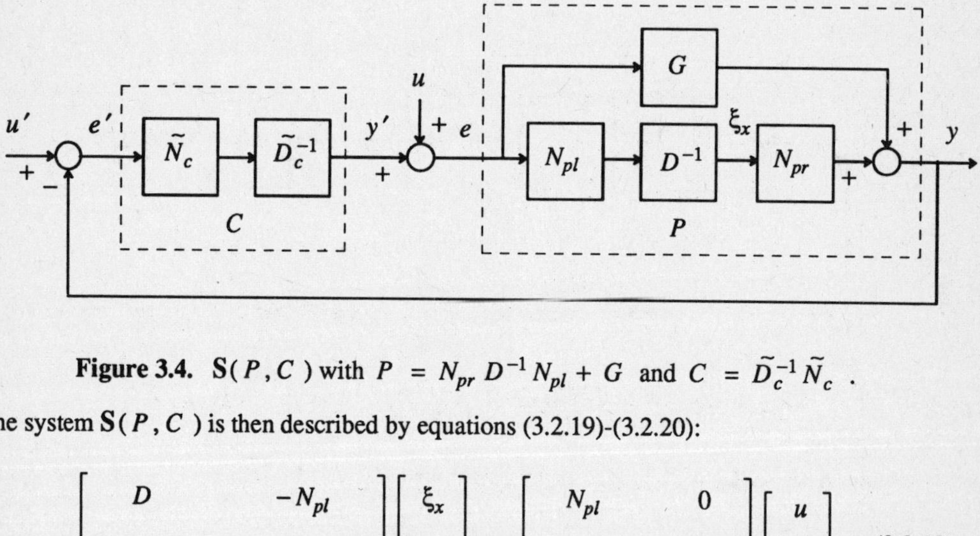

Figure 3.4. $\mathbf{S}(P, C)$ with $P = N_{pr} D^{-1} N_{pl} + G$ and $C = \tilde{D}_c^{-1} \tilde{N}_c$.

The system $\mathbf{S}(P, C)$ is then described by equations (3.2.19)-(3.2.20):

$$\begin{bmatrix} D & -N_{pl} \\ \tilde{N}_c N_{pr} & \tilde{D}_c + \tilde{N}_c G \end{bmatrix} \begin{bmatrix} \xi_x \\ y' \end{bmatrix} = \begin{bmatrix} N_{pl} & 0 \\ -\tilde{N}_c G & \tilde{N}_c \end{bmatrix} \begin{bmatrix} u \\ u' \end{bmatrix}, \quad (3.2.19)$$

$$\begin{bmatrix} N_{pr} & G \\ 0 & I_{n_i} \end{bmatrix} \begin{bmatrix} \xi_x \\ y' \end{bmatrix} = \begin{bmatrix} y \\ y' \end{bmatrix} - \begin{bmatrix} G & 0 \\ 0 & 0 \end{bmatrix} \begin{bmatrix} u \\ u' \end{bmatrix}. \quad (3.2.20)$$

Equations (3.2.19)-(3.2.20) are of the form

$$D_{H3}\,\xi_3 = N_{HL3}\,\bar{u}$$

$$N_{HR3}\,\xi_3 = \bar{y} - G_{H3}\,\bar{u} \quad .$$

As in Analysis 3.2.6 (i) above, it can be easily verified that $(N_{HR3}, D_{H3}, N_{HL3})$ is a b.c. triple and that $H_{\bar{y}\bar{u}} = N_{HR3}\,D_{H3}^{-1}\,N_{HL3} + G_{H3} \in \mathrm{m}(\mathrm{G})$ if and only if $\det D_{H3} \in \mathrm{I}$. Again by Assumption 3.2.1 (iii), $H_{\bar{y}\bar{u}} \in \mathrm{m}(\mathrm{G})$ and hence, $\det D_{H3} \in \mathrm{I}$. Consequently, $(N_{HR3}, D_{H3}, N_{HL3}, G_{H3})$ is a b.c.f.r. of $H_{\bar{y}\bar{u}}$ and hence, $\det D_{H3}$ is a characteristic determinant of $H_{\bar{y}\bar{u}}$.

(iv) Analysis of $\mathbf{S}(P, C)$ with $P = N_{pr}\,D^{-1}N_{pl} + G$ and $C = N_c\,D_c^{-1}$

Let (N_{pr}, D, N_{pl}, G) be any b.c.f.r. of $P \in \mathrm{m}(\mathrm{G})$ and let (N_c, D_c) be any r.c.f.r. of $C \in \mathrm{m}(\mathrm{G})$. The system $\mathbf{S}(P, C)$ in Figure 3.1 can be redrawn as in Figure 3.5 below, where $P = N_{pr}\,D^{-1}N_{pl} + G$ and $C = N_c\,D_c^{-1}$; note that $D\,\xi_x = N_{pl}\,e$, $y = N_{pr}\,\xi_x + G\,e$, where ξ_x denotes the *pseudo-state* of P and $D_c\,\xi_c = e'$, $y' = N_c\,\xi_c$, where ξ_c denotes the *pseudo-state* of C .

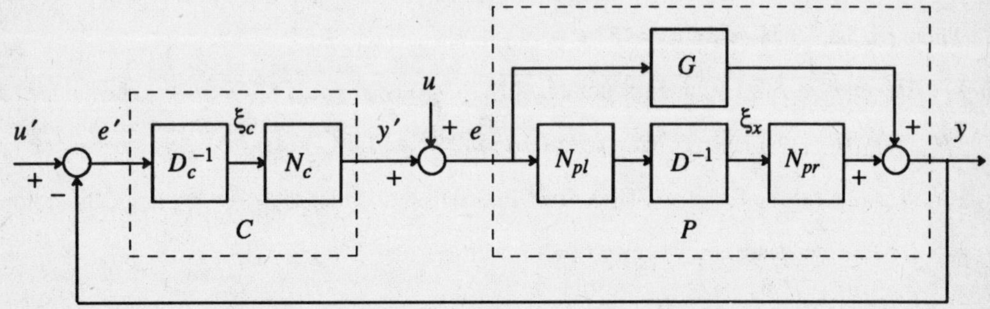

Figure 3.5. $\mathbf{S}(P, C)$ with $P = N_{pr}\,D^{-1}N_{pl} + G$ and $C = N_c\,D_c^{-1}$.

The system $\mathbf{S}(P, C)$ is then described by equations (3.2.21)-(3.2.22):

$$\begin{bmatrix} D & -N_{pl}\,N_c \\ N_{pr} & D_c + G\,N_c \end{bmatrix} \begin{bmatrix} \xi_x \\ \xi_c \end{bmatrix} = \begin{bmatrix} N_{pl} & 0 \\ -G & I_{n_o} \end{bmatrix} \begin{bmatrix} u \\ u' \end{bmatrix}, \tag{3.2.21}$$

$$\begin{bmatrix} N_{pr} & G\,N_c \\ 0 & N_c \end{bmatrix} \begin{bmatrix} \xi_x \\ \xi_c \end{bmatrix} = \begin{bmatrix} y \\ y' \end{bmatrix} - \begin{bmatrix} G & 0 \\ 0 & 0 \end{bmatrix} \begin{bmatrix} u \\ u' \end{bmatrix}. \tag{3.2.22}$$

Equations (3.2.21)-(3.2.22) are of the form

$$D_{H4}\,\xi_4 = N_{HL4}\,\bar{u}$$

$$N_{HR4}\,\xi_4 = \bar{y} - G_{H4}\,\bar{u} .$$

As in Analysis 3.2.6 (i) above, it can be easily verified that $(N_{HR4}, D_{H4}, N_{HL4})$ is a b.c. triple and that $H_{\bar{y}\bar{u}} = N_{HR4}\,D_{H4}^{-1}\,N_{HL4} + G_{H4} \in \mathrm{m(G)}$ if and only if $\det D_{H4} \in \mathrm{I}$. Again by Assumption 3.2.1 (iii), $H_{\bar{y}\bar{u}} \in \mathrm{m(G)}$ and hence, $\det D_{H4} \in \mathrm{I}$. Consequently, $(N_{HR4}, D_{H4}, N_{HL4}, G_{H4})$ is a b.c.f.r. of $H_{\bar{y}\bar{u}}$ and hence, $\det D_{H4}$ is a characteristic determinant of $H_{\bar{y}\bar{u}}$. □

Theorem 3.2.7. (H–stability of S(P, C))

Let Assumptions 3.2.1 (i) and (ii) hold; let (N_p, D_p) be any r.c.f.r., $(\tilde{D}_p, \tilde{N}_p)$ be any l.c.f.r., (N_{pr}, D, N_{pl}, G) be any b.c.f.r. over m(H) of $P \in \mathrm{m(G)}$; let $(\tilde{D}_c, \tilde{N}_c)$ be any l.c.f.r., (N_c, D_c) be any r.c.f.r. over m(H) of $C \in \mathrm{m(G)}$. Under these assumptions, the following five statements are equivalent:

(i) S(P, C) is H–stable ;

(ii) $D_{H1} := \left[\, \tilde{D}_c\,D_p + \tilde{N}_c\,N_p \,\right]$ is H–unimodular ; (3.2.23)

(iii) $D_{H2} := \left[\, \tilde{D}_p\,D_c + \tilde{N}_p\,N_c \,\right]$ is H–unimodular ; (3.2.24)

$$\textbf{(iv)}\quad D_{H3} := \begin{bmatrix} D & -N_{pl} \\ \tilde{N}_c\, N_{pr} & \tilde{D}_c + \tilde{N}_c\, G \end{bmatrix} \text{ is H–unimodular ;} \tag{3.2.25}$$

$$\textbf{(v)}\quad D_{H4} := \begin{bmatrix} D & -N_{pl}\, N_c \\ N_{pr} & D_c + G\, N_c \end{bmatrix} \text{ is H–unimodular .} \tag{3.2.26}$$

□

Note that each of statements (i) through (v) of Theorem 3.2.7 implies that the system $\mathbf{S}(P, C)$ is well-posed; consequently, we do not need to state a well-posedness assumption in the beginning of Theorem 3.2.7.

Proof

We prove the equivalence of statements **(i)** and **(ii)** of Theorem 3.2.7 using the system description in Analysis 3.2.6 (i): Suppose that $N_p\, D_p^{-1}$ is an r.c. factorization of P and $\tilde{D}_c^{-1}\, \tilde{N}_c$ is an l.c. factorization of C ; then $\mathbf{S}(P, C)$ is described by equations (3.2.14)-(3.2.15). If $\mathbf{S}(P, C)$ is H–stable, then $H_{\overline{yu}} \in \mathfrak{m}(\mathrm{H})$ and hence, condition (3.2.16) holds; equivalently, $\det D_{H1} \in \mathrm{I}$; therefore $(N_{HR1}, D_{H1}, N_{HL1}, G_{H1})$ is a b.c.f.r. of $H_{\overline{yu}}$. By Lemma 2.4.6, $H_{\overline{yu}} \in \mathfrak{m}(\mathrm{H})$ implies that $D_{H1}^{-1} \in \mathfrak{m}(\mathrm{H})$. Conversely, if condition (3.2.23) holds, then $D_{H1}^{-1} \in \mathfrak{m}(\mathrm{H})$ and hence, $H_{\overline{yu}} = N_{HR1}\, D_{H1}^{-1}\, N_{HL1} + G_{H1} \in \mathfrak{m}(\mathrm{H})$.

The equivalence of statement **(i)** to any of **(iii)**, **(iv)** or **(v)** follows similarly from Analysis 3.2.6 (ii), (iii) and (iv), respectively. □

Definition 3.2.8. (**H–stabilizing compensator** C)

(i) C is called an H–*stabilizing compensator for* P (abbreviated as: C H–*stabilizes* P) iff $C \in \mathrm{G}^{n_i \times n_o}$ and the system $\mathbf{S}(P, C)$ is H–stable.

(ii) The set

$$S(P) := \{ \ C \mid C \ \text{H–}\textit{stabilizes}\ P \ \}$$

is called the *set of all* H–*stabilizing compensators for* P in the system $\mathbf{S}(P, C)$.

Corollary 3.2.9

Let Assumption 3.2.1 (i) hold; let (N_p , D_p) be any r.c.f.r. and $(\tilde{D}_p , \tilde{N}_p)$ be any l.c.f.r. of $P \in \mathrm{m}(\mathrm{G})$. Under these assumptions, the following four statements are equivalent:

(i) C H–stabilizes P ;

(ii) an l.c.f.r. $(\tilde{D}_c , \tilde{N}_c)$ of $C \in \mathrm{m}(\mathrm{G})$ satisfies

$$\tilde{D}_c D_p + \tilde{N}_c N_p = I_{n_i} \; ; \tag{3.2.27}$$

(iii) an r.c.f.r. (N_c , D_c) of $C \in \mathrm{m}(\mathrm{G})$ satisfies

$$\tilde{N}_p N_c + \tilde{D}_p D_c = I_{n_o} \; ; \tag{3.2.28}$$

(iv) a doubly-coprime-fraction representation $((N_c , D_c) , (\tilde{D}_c , \tilde{N}_c))$ of $C \in \mathrm{m}(\mathrm{G})$ satisfies

$$\begin{bmatrix} \tilde{D}_c & \tilde{N}_c \\ -\tilde{N}_p & \tilde{D}_p \end{bmatrix} \begin{bmatrix} D_p & -N_c \\ N_p & D_c \end{bmatrix} = \begin{bmatrix} I_{n_i} & 0 \\ 0 & I_{n_o} \end{bmatrix} . \tag{3.2.29}$$

Proof

Suppose that statement (i) of Corollary 3.2.9 holds; then by Definition 3.2.8 (i), $C \in \mathrm{m}(\mathrm{G})$. Let $(\tilde{D}_c^* , \tilde{N}_c^*)$ be any l.c.f.r. and (D_c^* , N_c^*) be any r.c.f.r. of C ; then by Theorem 3.2.7, $\tilde{D}_c^* D_p + \tilde{N}_c^* N_p =: L \in \mathrm{m}(\mathrm{H})$ is H–unimodular. By Lemma 2.3.4 (ii), $(\tilde{D}_c , \tilde{N}_c) := (L^{-1} \tilde{D}_c^* , L^{-1} \tilde{N}_c^*)$ is also an l.c.f.r. of C ; but $(\tilde{D}_c , \tilde{N}_c)$ satisfies equation (3.2.27) since $(L^{-1} \tilde{D}_c^*) D_p + (L^{-1} \tilde{N}_c^*) N_p = I_{n_i}$, and hence, statement (ii) of Corollary 3.2.9 holds.

Now equation (3.2.27) implies that

$$\det(\tilde{D}_c D_p + \tilde{N}_c N_p) = 1 = \det \tilde{D}_c \det(I_{n_i} + C P) \det D_p . \tag{3.2.30}$$

By Lemma 2.4.4, $\det D_p \sim \det\tilde{D}_p$ and $\det\tilde{D}_c \sim \det D_c^*$; since $\det(I_{n_i} + C\,P) = \det(I_{n_o} + P\,C)$, equation (3.2.30) implies that

$$\det\tilde{D}_p \det(I_{n_o} + P\,C)\det D_c^* = \det(\tilde{D}_p D_c^* + \tilde{N}_p N_c^*) \sim 1 \qquad (3.2.31)$$

and therefore, $(\tilde{D}_p D_c^* + \tilde{N}_p N_c^*) =: R \in \mathfrak{m}(H)$ is H–unimodular. By Lemma 2.3.4 (i), $(N_c, D_c) := (D_c^* R^{-1}, N_c^* R^{-1})$ is also an r.c.f.r. of C ; but (N_c, D_c) satisfies equation (3.2.28) and hence, statement (iii) of Corollary 3.2.9 holds.

Since $\tilde{D}_p N_p = \tilde{N}_p D_p$ and $\tilde{D}_c N_c = \tilde{N}_c D_c$, statement (ii) also implies that the generalized Bezout identity (3.2.29) is satisfied and hence, statement (iv) holds.

Now suppose that C has an l.c.f.r. $(\tilde{D}_c, \tilde{N}_c)$ that satisfies equation (3.2.29), where (N_p, D_p) is any given r.c.f.r. of P ; then by Lemma 2.3.4, any other l.c.f.r. of C is of the form $(L\,\tilde{D}_c, L\,\tilde{N}_c)$ and any other r.c.f.r. of P is of the form $(N_p R, D_p R)$, where L, $R \in \mathfrak{m}(H)$ are H–unimodular matrices; therefore, condition (3.2.23) is satisfied for any l.c.f.r. of C and any r.c.f.r. of P since $D_{H1} = L\,R$, and hence, by Theorem 3.2.7, $S(P, C)$ is H–stable. □

Remark 3.2.10

Let (N_{pr}, D, N_{pl}, G) be a b.c.f.r. of $P \in \mathfrak{m}(G)$; let V_{pr}, U_{pr}, $\tilde{X}$, $\tilde{Y}$, $\tilde{V}$, $\tilde{U}$, V_{pl}, U_{pl}, X, Y, U, $V \in \mathfrak{m}(H)$ be as in the generalized Bezout identities (2.3.13)-(2.3.14), and let M_r, $M_l \in \mathfrak{m}(H)$ be the H–unimodular matrices defined in equations (2.3.15)-(2.3.16). Then the denominator matrix D_{H3} in equation (3.2.25) is H–unimodular if and only if

$$D_{H3} M_l = \begin{bmatrix} I_n & 0 \\ \tilde{N}_c (N_{pr} V_{pl} - G\,U_{pl}) - \tilde{D}_c U_{pl} & \tilde{N}_c (N_{pr} X + G\,Y) + \tilde{D}_c Y \end{bmatrix}$$

is H–unimodular, (3.2.32)

where $(\tilde{D}_c, \tilde{N}_c)$ is any l.c.f.r. of C. From (3.2.32), D_{H3} is H–unimodular if and only if

$$\tilde{D}_c Y + \tilde{N}_c (N_{pr} X + G Y) \quad \text{is H–unimodular}. \tag{3.2.33}$$

By Theorem 2.4.1, $(N_{pr} X + G Y, Y)$ is also an r.c.f.r. of P ; therefore condition (3.2.33) is equivalent to condition (3.2.23) of the H–stability Theorem 3.2.7.

Similarly, the denominator matrix D_{H4} in equation (3.2.26) is H–unimodular if and only if

$$M_r D_{H4} = \begin{bmatrix} I_n & (-V_{pr} N_{pl} + U_{pr} G) N_c + U_{pr} D_c \\ 0 & (\tilde{X} N_{pl} + \tilde{Y} G) N_c + \tilde{Y} D_c \end{bmatrix}$$

$$\text{is H–unimodular,} \tag{3.2.34}$$

where (N_c, D_c) is any r.c.f.r. of C . From (3.2.34), D_{H4} is H–unimodular if and only if

$$(\tilde{X} N_{pl} + \tilde{Y} G) N_c + \tilde{Y} D_c \quad \text{is H–unimodular}. \tag{3.2.35}$$

By Theorem 2.4.1, $(\tilde{Y}, \tilde{X} N_{pl} + \tilde{Y} G)$ is also an l.c.f.r. of P ; therefore condition (3.2.35) is equivalent to condition (3.2.24) of the H–stability Theorem 3.2.7.

By Corollary 3.2.9, C H–stabilizes P if and only if an l.c.f.r. $(\tilde{D}_c, \tilde{N}_c)$ of $C \in \mathrm{m}(\mathrm{G})$ satisfies

$$\tilde{D}_c Y + \tilde{N}_c (N_{pr} X + G Y) = I_{n_i} ; \tag{3.2.36}$$

equivalently, an r.c.f.r. (N_c, D_c) of $C \in \mathrm{m}(\mathrm{G})$ satisfies

$$(\tilde{X} N_{pl} + \tilde{Y} G) N_c + \tilde{Y} D_c = I_{n_o} . \tag{3.2.37}$$

□

We now parametrize the set $S(P)$ of all H–stabilizing compensators for P .

Theorem 3.2.11. (Parametrization of all H–stabilizing compensators in $\mathbf{S}(P, C)$)

Let Assumptions 3.2.1 (i) and (iii) hold; let (N_p, D_p) be any r.c.f.r. and $(\tilde{D}_p, \tilde{N}_p)$ be any l.c.f.r. of $P \in \mathbf{G}^{n_o \times n_i}$; let $V_p, U_p, \tilde{V}_p, \tilde{U}_p \in \mathrm{m}(\mathbf{H})$ be as in the generalized Bezout identity (2.3.12). Under these assumptions,

$$S(P) = \{ (V_p - Q\tilde{N}_p)^{-1}(U_p + Q\tilde{D}_p) \mid Q \in \mathbf{H}^{n_i \times n_o}, \det(V_p - Q\tilde{N}_p) \in \mathbf{I} \}; \tag{3.2.38}$$

equivalently,

$$S(P) = \{ (\tilde{U}_p + D_p Q)(\tilde{V}_p - N_p Q)^{-1} \mid Q \in \mathbf{H}^{n_i \times n_o}, \det(\tilde{V}_p - N_p Q) \in \mathbf{I} \}. \tag{3.2.39}$$

Furthermore, corresponding to each compensator $C \in S(P)$, there is a unique $Q \in \mathrm{m}(\mathbf{H})$ in the equivalent parametrizations (3.2.38) and (3.2.39). Equations (3.2.38) and (3.2.39) are bijections from $Q \in \mathrm{m}(\mathbf{H})$ to $C \in \hat{S}(\hat{P})$.

Remark 3.2.12. (H–stabilizing compensators based on a bicoprime factorization of P)

Let Assumptions 3.2.1 (i) and (iii) hold; let (N_{pr}, D, N_{pl}, G) be any b.c.f.r. of $P \in \mathrm{m}(\mathbf{G}_s)$; let $V_{pr}, U_{pr}, \tilde{X}, \tilde{Y}, \tilde{V}, \tilde{U}, V_{pl}, U_{pl}, X, Y, U, V \in \mathrm{m}(\mathbf{H})$ be as in the generalized Bezout identities (2.3.13)-(2.3.14). By Theorem 2.4.1, $(N_{pr} X + G Y, Y)$ is an r.c.f.r. and $(\tilde{Y}, \tilde{X} N_{pl} + \tilde{Y} G)$ is an l.c.f.r. of P; therefore, by Theorem 3.2.11, the set $S(P)$ of all H–stabilizing compensators is given by

$$S(P) = \{ (V + U V_{pr} N_{pl} - U U_{pr} G - Q(\tilde{X} N_{pl} + \tilde{Y} G))^{-1}(U U_{pr} + Q\tilde{Y}) \mid$$
$$Q \in \mathrm{m}(\mathbf{H}), \det(V + U V_{pr} N_{pl} - U U_{pr} G - Q(\tilde{X} N_{pl} + \tilde{Y} G)) \in \mathbf{I} \}; \tag{3.2.40}$$

equivalently,

$$S(P) = \{ (U_{pl}\tilde{U} + Y Q)(\tilde{V} + N_{pr} V_{pl}\tilde{U} - G U_{pl}\tilde{U} - (N_{pr} X + G Y)Q)^{-1} \mid$$
$$Q \in \mathrm{m}(\mathbf{H}), \det(\tilde{V} + N_{pr} V_{pl}\tilde{U} - G U_{pl}\tilde{U} - (N_{pr} X + G Y)Q) \in \mathbf{I} \}. \tag{3.2.41}$$

The equivalence of the representation (3.2.38) to (3.2.40) and of the representation (3.2.39) to (3.2.41) is easy to see by comparing the generalized Bezout identities (2.3.12) and (2.4.3).

Proof of Theorem 3.2.11

By Corollary 3.2.9, C is an H–stabilizing compensator for P if and only if an l.c.f.r. $(\tilde{D}_c, \tilde{N}_c)$ satisfies the Bezout identity (3.2.27); by Lemma 2.5.1, following Remark 2.5.2 (i), all solutions of (3.2.27) over $\mathfrak{m}(\mathrm{H})$ are given by equation (2.5.7). Similarly, C H–stabilizes P if and only if an r.c.f.r. (N_c, D_c) satisfies the Bezout identity (3.2.28); all solutions of (3.2.28) over $\mathfrak{m}(\mathrm{H})$ are given by equation (2.5.8).

Now if $C \in \mathfrak{m}(\mathrm{G})$ is an H–stabilizing compensator, then the denominator matrix $\tilde{D}_c$ is $(V_p - Q\tilde{N}_p)$ for some $Q \in \mathfrak{m}(\mathrm{H})$; by Definition 2.3.1 (vi), $\det(V_p - Q\tilde{N}_p) \in \mathrm{I}$. Conversely, if $Q \in \mathfrak{m}(\mathrm{H})$ is chosen so that $\det(V_p - Q\tilde{N}_p) \in \mathrm{I}$, then $(V_p - Q\tilde{N}_p)$ is a valid choice for the denominator matrix $\tilde{D}_c$. By Lemma 2.4.4, for all $Q \in \mathfrak{m}(\mathrm{H})$, $\det(V_p - Q\tilde{N}_p) \sim \det(\tilde{V}_p - N_p Q)$. We conclude then that the set of all H–stabilizing compensators is given by (3.2.38) and equivalently, by (3.2.39).

Now suppose that C is an H–stabilizing compensator; then from (3.2.38)-(3.2.39), an l.c.f.r. $(\tilde{D}_c, \tilde{N}_c)$ of C is given by

$$\begin{bmatrix} \tilde{D}_c & \tilde{N}_c \end{bmatrix} = \begin{bmatrix} I_{n_i} & Q_1 \end{bmatrix} \begin{bmatrix} V_p & U_p \\ -\tilde{N}_p & \tilde{D}_p \end{bmatrix} =: \begin{bmatrix} I_{n_i} & Q_1 \end{bmatrix} \bar{M} \,, \qquad (3.2.42)$$

for some $Q_1 \in \mathrm{H}^{n_i \times n_o}$; note that the matrix $\bar{M} \in \mathfrak{m}(\mathrm{H})$ is H–unimodular by the generalized Bezout identity (2.3.12); similarly, an r.c.f.r. (N_c, D_c) of C is given by

$$\begin{bmatrix} -N_c \\ D_c \end{bmatrix} = \begin{bmatrix} D_p & -\tilde{U}_p \\ N_p & \tilde{V}_p \end{bmatrix} \begin{bmatrix} -Q_2 \\ I_{n_o} \end{bmatrix} = \bar{M}^{-1} \begin{bmatrix} -Q_2 \\ I_{n_o} \end{bmatrix} , \qquad (3.2.43)$$

for some $Q_2 \in \mathrm{H}^{n_i \times n_o}$. But since $\tilde{D}_c^{-1} \tilde{N}_c = N_c D_c^{-1}$ implies that $\tilde{N}_c D_c - \tilde{D}_c N_c = 0$, from (3.2.42)-(3.2.43) and the generalized Bezout identity (2.3.12) we obtain

$$\begin{bmatrix} \tilde{D}_c & \tilde{N}_c \end{bmatrix} \bar{M}^{-1} \bar{M} \begin{bmatrix} -N_c \\ D_c \end{bmatrix} = \begin{bmatrix} I_{n_i} & Q_1 \end{bmatrix} \begin{bmatrix} -Q_2 \\ I_{n_o} \end{bmatrix} = Q_1 - Q_2 = 0 . \tag{3.2.44}$$

Therefore by (2.3.12) and by (3.2.44), we conclude that $(V_p - Q_1 \tilde{N}_p)^{-1} (U_p + Q_1 \tilde{D}_p) = (\tilde{U}_p + D_p Q_2) (\tilde{V}_p - N_p Q_2)^{-1}$ if and only if $Q_1 = Q_2$.

Now suppose that $C_1 \in S(P)$ has an l.c.f.r. $(\tilde{D}_{c1}, \tilde{N}_{c1})$ and $C_2 \in S(P)$ has an l.c.f.r. $(\tilde{D}_{c2}, \tilde{N}_{c2})$; by (3.2.38) and (3.2.42),

$$\begin{bmatrix} \tilde{D}_{c1} & \tilde{N}_{c1} \end{bmatrix} = \begin{bmatrix} I_{n_i} & Q_1 \end{bmatrix} \bar{M} = \tilde{D}_{c1} \begin{bmatrix} I_{n_i} & C_1 \end{bmatrix} , \tag{3.2.45}$$

for some $Q_1 \in \mathrm{H}^{n_o \times n_i}$ and

$$\begin{bmatrix} \tilde{D}_{c2} & \tilde{N}_{c2} \end{bmatrix} = \begin{bmatrix} I_{n_i} & Q_2 \end{bmatrix} \bar{M} = \tilde{D}_{c2} \begin{bmatrix} I_{n_i} & C_2 \end{bmatrix} , \tag{3.2.46}$$

for some $Q_2 \in \mathrm{H}^{n_o \times n_i}$. From (3.2.45)-(3.2.46), $C_1 = C_2$ if and only if

$$\tilde{D}_{c1}^{-1} \begin{bmatrix} I_{n_i} & Q_1 \end{bmatrix} \bar{M} = \tilde{D}_{c2}^{-1} \begin{bmatrix} I_{n_i} & Q_2 \end{bmatrix} \bar{M} ; \tag{3.2.47}$$

since $\bar{M}$ is H–unimodular, multiplying both sides of (3.2.47) by $\bar{M}^{-1} \in \mathrm{m(H)}$, we obtain $\tilde{D}_{c1}^{-1} = \tilde{D}_{c2}^{-1}$ and $\tilde{D}_{c1}^{-1} Q_1 = \tilde{D}_{c2}^{-1} Q_2$; i.e., $Q_1 = Q_2$. We conclude that $C_1 = C_2 \in S(P)$ if and only if $Q_1 = Q_2$. Consequently, there is a unique (matrix-) parameter $Q \in \mathrm{m(H)}$ corresponding to each H–stabilizing compensator $C \in S(P)$. □

Comment 3.2.13

(i) (Parametrization of all H–stabilizing compensators when $P \in \mathrm{m}(\mathrm{G}_S)$)

In addition to the assumptions of Theorem 3.2.11, suppose that $P \in \mathrm{m}(\mathrm{G}_S)$; then following Remark 2.5.2 (ii), $\det(V_p - Q \tilde{N}_p) \in \mathrm{I}$ and equivalently, $\det(\tilde{V}_p - N_p Q) \in \mathrm{I}$, *for all* $Q \in \mathrm{m}(\mathrm{H})$. Therefore, whenever $P \in \mathrm{m}(\mathrm{G}_S)$, $Q \in \mathrm{m}(\mathrm{H})$ is a *free* (matrix-) parameter in the equivalent parametrizations (3.2.38), (3.2.39), (3.2.40), (3.2.41). Hence $C = (V_p - Q \tilde{N}_p)^{-1} (U_p + Q \tilde{D}_p) = (\tilde{U}_p + D_p Q)(\tilde{V}_p - N_p Q)^{-1}$ H–stabilizes $P \in \mathrm{G}_S^{n_o \times n_i}$ for all choices of $Q \in \mathrm{H}^{n_i \times n_o}$; in particular,

$$C_o := V_p^{-1} U_p = \tilde{U}_p \tilde{V}_p^{-1} \tag{3.2.48}$$

is an H–stabilizing compensator when $P \in \mathrm{m}(\mathrm{G}_S)$. With $P = N_p D_p^{-1} = \tilde{D}_p^{-1} \tilde{N}_p$, Figures 3.6 and 3.7 show the H–stable system $\mathrm{S}(P, C)$, where $C = (V_p - Q \tilde{N}_p)^{-1} (U_p + Q \tilde{D}_p)$ and $C = (\tilde{U}_p + D_p Q)(\tilde{V}_p - N_p Q)^{-1}$, respectively.

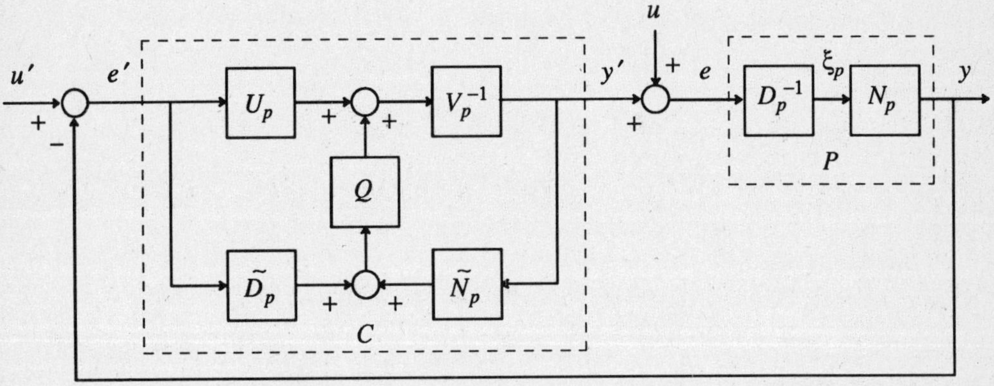

Figure 3.6. The system $\mathrm{S}(P, C)$,

with $P = N_p D_p^{-1} = \tilde{D}_p^{-1} \tilde{N}_p$, $C = (V_p - Q \tilde{N}_p)^{-1} (U_p + Q \tilde{D}_p)$.

If $P \notin \mathfrak{m}(\mathbf{G}_s)$, then $\det(V_p - Q\tilde{N}_p)$ and $\det(\tilde{V}_p - N_p Q)$ are not necessarily in $\mathbf{I}$ for all $Q \in \mathfrak{m}(\mathbf{H})$; in particular, $\det V_p$ and $\det\tilde{V}_p$ are not necessarily in $\mathbf{I}$ for any V_p and $\tilde{V}_p$ that satisfy the generalized Bezout identity (2.3.12). In Figures 3.6 and 3.7, Q should then be restricted to those $\mathbf{H}$–stable matrices for which condition (2.5.9) holds; V_p and $\tilde{V}_p$ should be replaced by $(V_p - Q^o\tilde{N}_p)$ and $(\tilde{V}_p - N_p Q^o)$, respectively, where $Q^o \in \mathfrak{m}(\mathbf{H})$ is chosen so that these are valid denominator matrices for the compensator. Note that following Remark 2.5.2 (ii), if the (matrix-) parameter $Q \in \mathfrak{m}(\mathbf{H})$ in the representations (3.2.38)-(3.2.39) is chosen as $Q^o \in \mathfrak{m}(\mathbf{H})$ given by (2.5.13), then $\det(V_p - Q^o\tilde{N}_p) \in \mathbf{I}$ and equivalently, $\det(\tilde{V}_p - N_p Q^o) \in \mathbf{I}$.

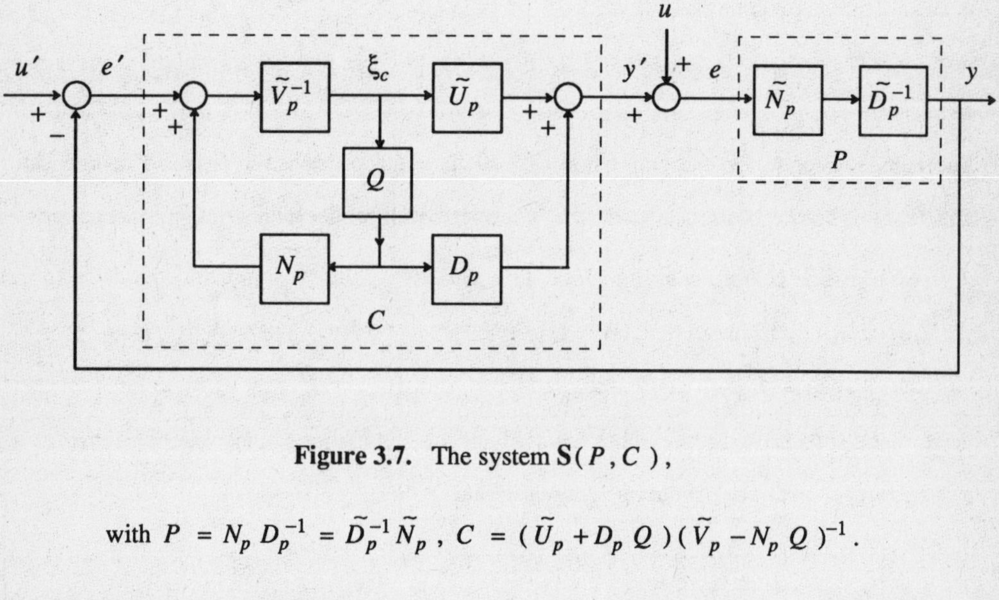

Figure 3.7. The system $\mathbf{S}(P, C)$,

with $P = N_p D_p^{-1} = \tilde{D}_p^{-1}\tilde{N}_p$, $C = (\tilde{U}_p + D_p Q)(\tilde{V}_p - N_p Q)^{-1}$.

(ii) (Observer-based compensator)

Suppose that $\mathbf{S}(P, C)$ is a lumped-parameter, continuous-time, linear, time-invariant system; let $P \in \mathfrak{m}(\mathbb{R}_p(s))$ be represented by its state-space representation $(\bar{A}, \bar{B}, \bar{C}, \bar{E})$ as in Example 2.4.3. An r.c.f.r. (N_p, D_p) and an l.c.f.r. $(\tilde{D}_p, \tilde{N}_p)$ of P are then given by (2.4.18) and (2.4.19), respectively. Let

$$V_p := I_{n_i} + K A_f (\bar{B} - F\bar{E}) \quad , \quad U_p := K A_f F \quad ,$$

$$\tilde{V}_p := I_{n_o} + (C - \bar{E} K) A_k F \quad , \quad \tilde{U}_p := K A_k F \ ;$$

by the generalized Bezout identity (2.4.17), $V_p D_p + U_p N_p = I_{n_i}$ and $\tilde{D}_p \tilde{V}_p + \tilde{N}_p \tilde{U}_p = I_{n_o}$. Now since the matrices $A_k := (s I_n - \bar{A} + \bar{B} K)^{-1}$, $A_f := (s I_n - \bar{A} + F \bar{C})^{-1} \in \mathrm{m}(\mathrm{R_U})$, defined in equation (2.4.14), are strictly proper, $\det V_p = \det(I_{n_i} + K A_f (\bar{B} - F \bar{E})) \in \mathrm{I}$ and $\det \tilde{V}_p = \det(I_{n_o} + (\bar{C} - \bar{E} K) A_k F) \in \mathrm{I}$. Since V_p^{-1}, $\tilde{V}_p^{-1} \in \mathrm{m}(\mathbb{R}_p(s))$, V_p and $\tilde{V}_p$ are already valid compensator denominator matrices in this case for any $P \in \mathrm{m}(\mathbb{R}_p(s))$; therefore $C_o := V_p^{-1} U_p = \tilde{U}_p \tilde{V}_p^{-1} \in \mathrm{m}(\mathbb{R}_{sp}(s))$ is an H–stabilizing compensator, where

$$C_o = (I_{n_i} + K A_f (\bar{B} - F \bar{E}))^{-1} K A_f F = K A_k F (I_{n_o} + (C - \bar{E} K) A_k F)^{-1}$$

$$= K (s I_n - \bar{A} + \bar{B} K + F \bar{C} - F \bar{E} K)^{-1} F \ . \tag{3.2.49}$$

The compensator C_o in the expression (3.2.49) is a full-order observer-based compensator; note that this compensator is always strictly proper for all proper (and strictly proper) plants.

Now $(V_p - Q \tilde{N}_p)$ and $(\tilde{V}_p - N_p Q)$ are also valid denominator matrices for all $Q \in \mathrm{m}(\mathrm{R_U})$ such that $\det(V_p(\infty) - Q(\infty) \tilde{N}_p(\infty)) \sim \det(\tilde{V}_p(\infty) - N_p(\infty) Q(\infty)) \neq 0$; note that $V_p(\infty) = I_{n_i}$, $\tilde{V}_p(\infty) = I_{n_o}$, $N_p(\infty) = \tilde{N}_p(\infty) = \bar{E}$; therefore in the representations (3.2.38)-(3.2.39) of the set $S(P)$ of all H–stabilizing compensators, the (matrix-) parameter $Q \in \mathrm{m}(\mathrm{R_U})$ should be chosen so that

$$\det(\tilde{V}_p(\infty) - N_p(\infty) Q(\infty)) = \det(I_{n_i} - Q(\infty) \bar{E})$$
$$= \det(I_{n_o} - \bar{E} Q(\infty)) = \det(V_p(\infty) - Q(\infty) \tilde{N}_p(\infty)) \neq 0 \ . \tag{3.2.50}$$

Note that condition (3.2.50) is automatically satisfied for all $Q \in \mathrm{m}(\mathrm{R_U}) \cap \mathrm{m}(\mathbb{R}_{sp}(s))$. Figure 3.8 shows the system $\mathrm{S}(P, C)$, where the plant transfer function is obtained from its state-space representation $(\bar{A}, \bar{B}, \bar{C}, \bar{E})$ as $P = \bar{C}(s I_n - \bar{A})^{-1} \bar{B} + \bar{E}$. If the parameter Q is chosen as a real constant matrix such that $\det(I_{n_i} - Q \bar{E}) \neq 0$, then $\tilde{x}$ is the state of the compensator C in Figure 3.8; in this case, the state-space representation $(\bar{A}_c, \bar{B}_c, \bar{C}_c, \bar{E}_c)$ is given by:

$$\bar{A}_c = \bar{A} - F\bar{C} + (\bar{B} - F\bar{E})(I_{n_i} - Q\bar{E})^{-1}(-K + Q\bar{C}) ,$$

$$\bar{B}_c = (\bar{B} - F\bar{E})(I_{n_i} - Q\bar{E})^{-1}Q - F ,$$

$$\bar{C}_c = (I_{n_i} - Q\bar{E})^{-1}(-K + Q\bar{C}) , \quad \bar{E}_c = (I_{n_i} - Q\bar{E})^{-1}Q , \tag{3.2.51}$$

where $Q \in \mathfrak{m}(\mathbb{R})$ is such that condition (3.2.50) holds, i.e., $(I_{n_i} - Q\bar{E})^{-1} \in \mathfrak{m}(\mathbb{R})$.

The compensator transfer function is obtained from Figure 3.8 as:

$$C = (I_{n_i} - Q\bar{E})^{-1}(-K + Q\bar{C})\left[A_f^{-1} - (\bar{B} - F\bar{E})(I_{n_i} - Q\bar{E})^{-1}(-K + Q\bar{C})\right]^{-1}$$

$$\cdot [(\bar{B} - F\bar{E})(I_{n_i} - Q\bar{E})^{-1}Q - F] + (I_{n_i} - Q\bar{E})^{-1}Q . \tag{3.2.52}$$

The compensator transfer function can be obtained also using the pseudo-state ξ_c : From Figure 3.8,

$$\tilde{x} = A_k(\bar{B}Q - F)\xi_c ,$$

$$\left[I_{n_o} + (\bar{C} - \bar{E}K)A_k F - (\bar{C}A_k\bar{B} + \bar{E}(I_{n_i} - KA_k\bar{B}))Q\right]\xi_c = e' ,$$

$$\left[KA_kF + (I_{n_i} - KA_kB)Q\right]\xi_c = y' .$$

Therefore,

$$C = \left[KA_kF + (I_{n_i} - KA_kB)Q\right]$$

$$\cdot\left[I_{n_o} + (\bar{C} - \bar{E}K)A_kF - (\bar{C}A_k\bar{B} + \bar{E}(I_{n_i} - KA_k\bar{B}))Q\right]^{-1} . \tag{3.2.53}$$

Note that this is in the right-coprime factorization form $(\tilde{U}_p + D_pQ)(\tilde{V}_p - N_pQ)^{-1}$ as in (3.2.39); to obtain the left-coprime factorization form as in (3.2.38), we write

$$\left[I_{n_i} + KA_f(\bar{B} - F\bar{E}) - Q(\bar{C}A_f\bar{B} + (I_{n_o} - \bar{C}A_fF)\bar{E})\right]y'$$

$$= \left[K\, A_f\, F + Q\, (I_{n_o} - \bar{C}\, A_f\, F\,) \right] e' \ .$$

Therefore,

$$C = \left[I_{n_i} + K\, A_f\, (\bar{B} - F\, \bar{E}\,) - Q\, (\bar{C}\, A_f\, \bar{B} + (I_{n_o} - \bar{C}\, A_f\, F\,)\bar{E}\,) \right]^{-1}$$

$$\cdot \left[K\, A_f\, F + Q\, (I_{n_o} - \bar{C}\, A_f\, F\,) \right] . \qquad (3.2.54)$$

In equations (3.2.52), (3.2.53) and (3.2.54) representing the compensator transfer function C, the (matrix-) parameter $Q \in \mathrm{m}(\mathrm{R_U})$ is chosen so that condition (3.2.50) holds, i.e., $\det(I_{n_i} - Q(\infty)\,\bar{E}\,) \neq 0$. Following Remark 2.5.2 (iv), the H–stabilizing compensator C in (3.2.52) (and equivalently in (3.2.53) and (3.2.54)) is strictly proper if and only if $Q \in \mathrm{m}(\mathrm{H}) \cap \mathrm{m}(\mathbb{R}_{sp}(s))$ since equation (2.5.11) is satisfied if and only if $Q(\infty) = -U_p(\infty)\,\tilde{D}_p^{-1}(\infty) = 0$; this can also be seen from (3.2.53) and (3.2.54) since $C(\infty) = (I_{n_i} - Q(\infty)\,\bar{E}\,)^{-1}\, Q(\infty)$. Note that by equation (3.2.5), condition (3.2.50) is equivalent to the well-posedness of the system $\mathbf{S}(P, C\,)$ since $\det(I_{n_i} - Q(\infty)\,\bar{E}\,) = \det(I_{n_i} + C(\infty)\,P(\infty)\,) = \det(I_{n_o} + P(\infty)\,C(\infty)\,)$, where C is given by the equivalent representations (3.2.52), (3.2.53) and (3.2.54) and $P(\infty) = \bar{E}$.

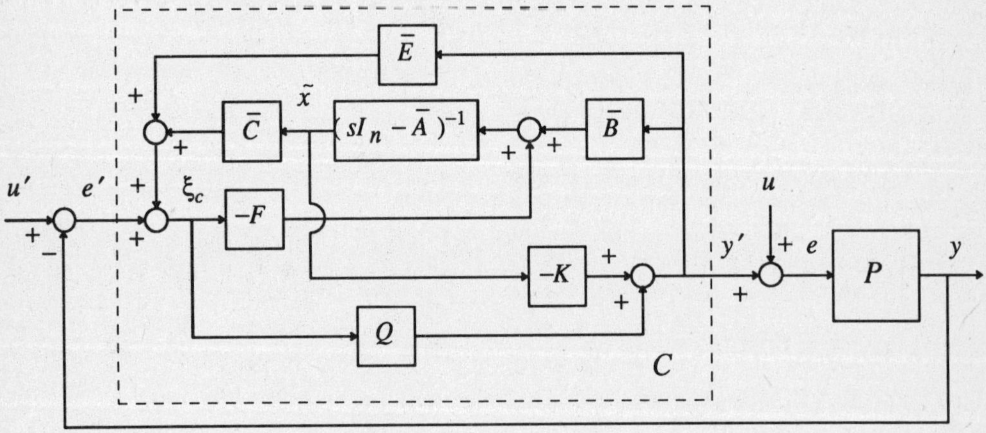

Figure 3.8. The system $\mathbf{S}(P, C\,)$, with an observer-based compensator.

(iii) (Parametrization of all H–stabilizing compensators when the plant is H–stable)

In addition to the assumptions of Theorem 3.2.11, suppose that $P \in \mathrm{m}(\mathrm{H})$; then following Remark 2.5.2 (v), we can choose (P , I_{n_i}) as an r.c.f.r. and (I_{n_o} , P) as an l.c.f.r. of P. In this case, from (3.2.38)-(3.2.39), the set $S (P)$ of all H–stabilizing compensators is given by

$$S(P) = \{ \, (I_{n_i} - Q\,P\,)^{-1} Q = Q\,(I_{n_o} - P\,Q\,)^{-1} \; \big| \; Q \in \mathrm{H}^{n_i \times n_o} \, ,$$

$$\det(I_{n_i} - Q\,P\,) = \det(I_{n_o} - P\,Q\,) \in \mathrm{I} \; \} \, . \tag{3.2.55}$$

If $P \in \mathrm{m}(\mathrm{H}) \cap \mathrm{m}(\mathrm{G}_s)$, then following Comment 3.2.13 (i) above, $\det(I_{n_i} - Q\,P\,) = \det(I_{n_o} - P\,Q\,) \in \mathrm{I}$, *for all* $Q \in \mathrm{m}(\mathrm{H})$.

(iv) (Parametrization of all strictly proper compensators)

Suppose that H is the ring $\mathrm{R_U}$ as in Section 2.2; let $P \in \mathrm{m}(\mathbb{R}_p(s))$; then following Remark 2.5.2 (iv), the set of all strictly proper compensators $C \in \mathrm{m}(\mathbb{R}_{sp}(s))$ is given by the equivalent representations (3.2.56) and (3.2.57) below:

$$\{ \, (V_p - Q\,\tilde{N}_p\,)^{-1} (U_p + Q\,\tilde{D}_p\,) \; \big| \; Q \in \mathrm{m}(\mathrm{R_U}) \, , \; Q(\infty) = -U_p(\infty)\,\tilde{D}_p^{-1}(\infty) \; \} \, , \tag{3.2.56}$$

$$\{ \, (\tilde{U}_p + D_p\,Q\,)(\tilde{V}_p - N_p\,Q\,)^{-1} \; \big| \; Q \in \mathrm{m}(\mathrm{R_U}) \, , \; Q(\infty) = -D_p^{-1}(\infty)\,\tilde{U}_p(\infty) \; \} \, ; \tag{3.2.57}$$

note that $D_p\,U_p = \tilde{U}_p\,\tilde{D}_p$ by the generalized Bezout identity (2.3.12). Choosing $Q \in \mathrm{m}(\mathrm{R_U})$ such that $Q(\infty) = -U_p(\infty)\,\tilde{D}_p^{-1}(\infty) = -D_p^{-1}(\infty)\,\tilde{U}_p(\infty)$ as in equation (2.5.14) is a sufficient condition for $\det(V_p - Q\,\tilde{N}_p\,) \in \mathrm{I}$ and equivalently, $\det(\tilde{V}_p - N_p\,Q\,) \in \mathrm{I}$.

(v) (Parametrization of all plants such that $\mathrm{S}(P, C)$ is H–stable)

The conditions for H–stability of the system $\mathrm{S}(P, C)$ are symmetric in the plant P and the compensator C. Suppose that Assumptions 3.2.1 (ii) and (iii) hold, where $C \in \mathrm{m}(\mathrm{G})$ is given and $(\tilde{D}_c , \tilde{N}_c)$ is an l.c.f.r., (N_c, D_c) is an r.c.f.r of C; let V_c, U_c, $\tilde{V}_c$, $\tilde{U}_c \in$

$\mathfrak{m}(\mathrm{H})$ be as in the generalized Bezout identity (2.5.19). Under these assumptions, $\mathrm{S}(P, C)$ is H–stable with $P \in \mathfrak{m}(\mathrm{G})$ if and only if P is of the form

$$(\tilde{U}_c + D_c Q_p)(\tilde{V}_c - N_c Q_p)^{-1} = (V_c - Q_p \tilde{N}_c)^{-1}(U_c + Q_p \tilde{D}_c) \quad (3.2.58)$$

for some $Q_p \in \mathfrak{m}(\mathrm{H})$ such that $\det(\tilde{V}_c - N_c Q_p) \sim \det(V_c - Q_p \tilde{N}_c) \in \mathrm{I}$. □

3.2.14. Achievable input-output maps of $\mathrm{S}(P, C)$

An important consequence of the parametrization of all H–stabilizing compensators is that the set of all achievable closed-loop I/O maps $H_{\overline{yu}} : \begin{bmatrix} u \\ u' \end{bmatrix} \mapsto \begin{bmatrix} y \\ y' \end{bmatrix}$ can now be described explicitly using $\mathcal{S}(P)$ given in the equivalent representations (3.2.38) and (3.2.39).

The set

$$\mathrm{A}(P) := \{ H_{\overline{yu}} \mid C \text{ H–stabilizes } P \}$$

is called *the set of all achievable I/O maps* of the unity-feedback system $\mathrm{S}(P, C)$.

By Theorem 3.2.11, $\mathrm{A}(P) = \{ H_{\overline{yu}} \mid C \in \mathcal{S}(P) \}$, where $\mathcal{S}(P)$ is the set of all H–stabilizing compensators given by (3.2.38) and equivalently, (3.2.39). We obtain the set of all achievable I/O maps from the expression (3.2.6) for $H_{\overline{yu}}$ by substituting an r.c. factorization $N_p D_p^{-1}$ for P and an l.c. factorization $\tilde{D}_c^{-1} \tilde{N}_c$ for C; from (3.2.38), $\tilde{D}_c^{-1} \tilde{N}_c =$ $(V_p - Q \tilde{N}_p)^{-1}(U_p + Q \tilde{D}_p)$ for some $Q \in \mathfrak{m}(\mathrm{H})$ such that $\det(V_p - Q \tilde{N}_p) \in \mathrm{I}$, where $V_p, U_p \in \mathfrak{m}(\mathrm{H})$ as in the generalized Bezout identity (2.3.12):

$$\mathrm{A}(P) = \left\{ H_{\overline{yu}} = \begin{bmatrix} N_p (V_p - Q \tilde{N}_p) & N_p (U_p + Q \tilde{D}_p) \\ D_p (V_p - Q \tilde{N}_p) - I_{n_i} & D_p (U_p + Q \tilde{D}_p) \end{bmatrix} \;\middle|\; Q \in \mathfrak{m}(\mathrm{H}),\ \det(V_p - Q \tilde{N}_p) \in \mathrm{I} \right\}. \quad (3.2.59)$$

The set $\mathrm{A}(P)$ of all achievable I/O maps can also be described as in (3.2.60) below, where an l.c. factorization $\tilde{D}_p^{-1} \tilde{N}_p$ for P and an r.c. factorization $N_c D_c^{-1}$ for C are substituted in the expression (3.2.7) for $H_{\overline{yu}}$; from (3.2.39), $N_c D_c^{-1} = (\tilde{U}_p + D_p Q)(\tilde{V}_p - N_p Q)^{-1}$ for some $Q \in \mathrm{m}(\mathrm{H})$ such that $\det(\tilde{V}_p - N_p Q) \in \mathrm{I}$, where $\tilde{U}_p, \tilde{V}_p \in \mathrm{m}(\mathrm{H})$ are as in the generalized Bezout identity (2.3.12):

$$\mathrm{A}(P) = \left\{ H_{\overline{yu}} = \begin{bmatrix} (\tilde{V}_p - N_p Q)\tilde{N}_p & I_{n_o} - (\tilde{V}_p - N_p Q)\tilde{D}_p \\ -(\tilde{U}_p + D_p Q)\tilde{N}_p & (\tilde{U}_p + D_p Q)\tilde{D}_p \end{bmatrix} \;\middle|\; Q \in \mathrm{m}(\mathrm{H}),\ \det(\tilde{V}_p - N_p Q) \in \mathrm{I} \right\}. \tag{3.2.60}$$

The representation (3.2.59) and equivalently, (3.2.60), is a parametrization of all closed-loop I/O maps. Each of the four I/O maps of $H_{\overline{yu}}$ in the equivalent representations (3.2.59)-(3.2.60) is affine in the (matrix-) parameter Q ; the matrix $Q \in \mathrm{m}(\mathrm{H})$ is a design parameter. Since each achievable closed-loop map depends on the same (matrix-) parameter Q , the system $\mathrm{S}(P, C)$ is called a *one-parameter* design or a *one-degree-of-freedom* compensation scheme.

Now suppose that the plant P is H–stable; following Comment 3.2.13 (iii), the set $\mathrm{A}(P)$ of all achievable I/O maps is given by

$$\mathrm{A}(P) = \left\{ H_{\overline{yu}} = \begin{bmatrix} P(I_{n_i} - QP) & PQ \\ -QP & Q \end{bmatrix} \;\middle|\; Q \in \mathrm{m}(\mathrm{H}),\ \det(I_{n_i} - QP) = \det(I_{n_o} - PQ) \in \mathrm{I} \right\}. \tag{3.2.61}$$

From the representation (3.2.59) we see that the plant output $y = N_p \begin{bmatrix} (V_p - Q\tilde{N}_p) & (U_p + Q\tilde{D}_p) \end{bmatrix} \begin{bmatrix} u \\ u' \end{bmatrix}$; hence, for all $C \in \mathrm{m}(\mathrm{H})$, the achievable I/O maps H_{yu} and $H_{yu'}$ both have the plant numerator matrix N_p as a left factor. This implies that the plant dynamics impose a constraint on the I/O maps H_{yu} and $H_{yu'}$.

3.2.15. Decoupling in $\mathbf{S}(P, C)$

Consider the unity-feedback system $\mathbf{S}(P, C)$; let Assumptions 3.2.1 hold. We now consider the problem of decoupling in $\mathbf{S}(P, C)$.

Let (N_p, D_p) be an r.c.f.r. of P and $(\tilde{D}_p, \tilde{N}_p)$ be an l.c.f.r. of P. Let V_p, U_p, $\tilde{V}_p$, $\tilde{U}_p$ be as in the generalized Bezout identity (2.3.12).

The system $\mathbf{S}(P, C)$ is said to be *decoupled* iff $\mathbf{S}(P, C)$ is H–stable and the closed-loop map $H_{yu'} : u' \mapsto y$ is *diagonal* and *nonsingular.*

A compensator C is said to decouple the system $\mathbf{S}(P, C)$ iff $C \in \mathcal{S}(P)$ and the map $H_{yu'} : u' \mapsto y$ is *diagonal* and *nonsingular.*

By equation (3.2.59), $H_{yu'}$ is an achievable map of $\mathbf{S}(P, C)$ if and only if

$$H_{yu'} = N_p \left(U_p + Q \tilde{D}_p \right) \tag{3.2.62}$$

for some $Q \in \mathrm{H}^{n_i \times n_o}$ such that $\det(V_p - Q \tilde{N}_p) \in \mathrm{I}$. There exists an H–stabilizing compensator that decouples $\mathbf{S}(P, C)$ if and only if there exists a $Q \in \mathrm{m}(\mathrm{H})$ such that $\det(V_p - Q \tilde{N}_p) \in \mathrm{I}$ for which the map $H_{yu'}$ in (3.2.62) is diagonal and nonsingular.

Let *rank* P denote the normal rank of $P \in \mathrm{G}^{n_o \times n_i}$ (i.e., the rank of P over the ring G). Since $P = N_p D_p^{-1}$, $rank\ P \le rank\ N_p$; since $N_p = P D_p$, $rank\ N_p \le rank\ P$; therefore, $rank\ N_p = rank\ P$ and similarly, $rank\ \tilde{N}_p = rank\ P$. By equation (3.2.62), if $H_{yu'}$ is an achievable map of $\mathbf{S}(P, C)$, then $rank\ H_{yu'} \le rank\ N_p$; if the system $\mathbf{S}(P, C)$ is actually decoupled, then $H_{yu'}$ is nonsingular and hence, $rank\ H_{yu'} = n_o$. But since $N_p \in \mathrm{H}^{n_o \times n_i}$, $n_o = rank\ H_{yu'} \le rank\ N_p \le n_o$ implies that, if decoupling is achieved, then $rank\ N_p = n_o = rank\ P$. Hence, a necessary condition for decoupling is that

$$rank\ P = n_o \le n_i \ . \tag{3.2.63}$$

In the case that P ˜ is square, (i.e., N_p and $\tilde{N}_p$ are also square), condition (3.2.63) means that $\det P \neq 0$ (i.e., $\det N_p \neq 0$ and $\det \tilde{N}_p \neq 0$ because $\det D_p \sim \det \tilde{D}_p \in \mathrm{I}$ by Definition 2.3.1).

In the rest of this section suppose that condition (3.2.63) holds and that the plant P is H–stable; in this case, it is always possible to find a $C \in \mathcal{S}(P)$ that decouples $\mathbf{S}(P, C)$: From (3.2.61), $H_{yu'}$ is achievable if and only if

$$H_{yu'} = P\,Q \tag{3.2.64}$$

for some $Q \in \mathrm{m}(\mathrm{H})$ such that $\det(I_{n_i} - Q\,P) \in \mathrm{I}$. For $k = 1, \cdots, n_o$, let $\Delta_{Lk} \in \mathrm{H}$ be a greatest-common-divisor (g.c.d.) of the entries in the k-th row of P. Let

$$\Delta_L := diag\begin{bmatrix} \Delta_{L1} & \cdots & \Delta_{Lno} \end{bmatrix}; \tag{3.2.65}$$

the diagonal matrix $\Delta_L \in \mathrm{m}(\mathrm{H})$ is nonsingular since $\Delta_{Lk} \neq 0$ for $k = 1, \cdots, n_o$; furthermore, the diagonal entries Δ_{Lk} of Δ_L are unique within factors in J. Let

$$P =: \Delta_L\,\tilde{P}\;; \tag{3.2.66}$$

clearly, $rank\ P \leq rank \tilde{P}$; since $\det\Delta_L \neq 0$, $rank\ \tilde{P} = rank\,(\Delta_L{}^{-1} P) \leq rank\ P$; therefore, condition (3.2.63) implies that $rank\ \tilde{P} = n_o$ and hence, $\tilde{P} \in \mathrm{H}^{n_o \times n_i}$ has a right-inverse denoted by $\tilde{P}^I$. Note that $\tilde{P}^I \in \mathrm{m}(\mathrm{F})$. Write the ij-th entry of $\tilde{P}^I$ as $\frac{m_{ij}}{d_{ij}}$, where m_{ij}, $d_{ij} \in \mathrm{H}$, $d_{ij} \neq 0$ and (m_{ij}, d_{ij}) is a coprime pair over H. For $j = 1, \cdots, n_o$, let $\Delta_{Rj} \in \mathrm{H}$ be a least-common-multiple (l.c.m.) of the denominators of the entries in the j-th column of $\tilde{P}^I$. Let

$$\Delta_R := diag\begin{bmatrix} \Delta_{R1} & \cdots & \Delta_{Rno} \end{bmatrix}; \tag{3.2.67}$$

the diagonal matrix $\Delta_R \in \mathrm{m}(\mathrm{H})$ is nonsingular since $\Delta_{Rj} \neq 0$ for $j = 1, \cdots, n_o$; furthermore, the diagonal entries Δ_{Rj} of Δ_R are unique within factors in J. (In the case that $\tilde{P}^I \in \mathrm{m}(\mathrm{H})$, the denominators $d_{ij} \in \mathrm{J}$ and hence, without loss of generality, Δ_R is the identity matrix I_{n_o}.) Now by definition, $\Delta_{Rj} = b_{ij}\,d_{ij}$ for some $b_{ij} \in \mathrm{H}$; therefore the ij-th entry of $\tilde{P}^I \Delta_R$ is $\frac{m_{ij}}{d_{ij}} \Delta_{Rj} = m_{ij}\,b_{ij} \in \mathrm{H}$; hence,

$$\tilde{P}^I \Delta_R \in \mathrm{m}(\mathrm{H})\,. \tag{3.2.68}$$

If we choose the (matrix-) parameter $Q \in \mathrm{m}(\mathrm{H})$ as

$$Q := \tilde{P}^{I} \Delta_R Q_d \in \mathrm{m}(\mathrm{H}) , \tag{3.2.69}$$

where $Q_d \in \mathrm{m}(\mathrm{H})$ is diagonal and nonsingular, then by (3.2.68), Q is H–stable. The compensator $C = (I_{n_i} - Q P)^{-1} Q = Q (I_{n_o} - P Q)^{-1}$ decouples $\mathrm{S}(P, C)$, where Q is chosen as in (3.2.69) and $Q_d \in \mathrm{m}(\mathrm{H})$ is a diagonal nonsingular matrix such that

$$\det(I_{n_i} - Q P) = \det(I_{n_o} - P Q) = \det(I_{n_o} - \Delta_L \Delta_R Q_d) \in \mathrm{I} . \tag{3.2.70}$$

Condition (3.2.70) is automatically satisfied for all $Q_d \in \mathrm{m}(\mathrm{H}) \cap \mathrm{m}(\mathrm{G}_s)$. If $P \in \mathrm{m}(\mathrm{H}) \cap \mathrm{m}(\mathrm{G}_s)$, then $\Delta_L \in \mathrm{m}(\mathrm{H}) \cap \mathrm{m}(\mathrm{G}_s)$; in this case, condition (3.2.70) is satisfied *for all* diagonal, nonsingular $Q_d \in \mathrm{m}(\mathrm{H})$. The diagonal map achieved is given by

$$H_{yu'} = \Delta_L \Delta_R Q_d . \tag{3.2.71}$$

If the map $H_{yu'}$ is required to be *block-diagonal*, then one way to achieve this is to choose $Q_d \in \mathrm{m}(\mathrm{H})$ block-diagonal, where Q_d satisfies condition (3.2.70).

Theorem 3.2.16 below summarizes decoupling in the system $\mathrm{S}(P, C)$ when the plant P is H–stable:

Theorem 3.2.16. (Class of all achievable diagonal $H_{yu'}$)

Let $P \in \mathrm{H}^{n_o \times n_i}$ and let *rank* $P = n_o$; then the set of all compensators that decouple the system $\mathrm{S}(P, C)$ is given by

$$\{ \tilde{P}^{I} \Delta_R Q_d (I_{n_o} - \Delta_L \Delta_R Q_d)^{-1} \mid Q_d \in \mathrm{H}^{n_o \times n_o} \text{ is diagonal, nonsingular}$$
$$\text{and } \det(I_{n_o} - \Delta_L \Delta_R Q_d) \in \mathrm{I} \} . \tag{3.2.72}$$

Furthermore, the set of all achievable, diagonal, nonsingular maps $H_{yu'} : u' \mapsto y$ is given by

$$\{ \Delta_L \Delta_R Q_d \mid Q_d \in \mathrm{H}^{n_o \times n_o} \text{ is diagonal, nonsingular}$$
$$\text{and } \det(I_{n_o} - \Delta_L \Delta_R Q_d) \in \mathrm{I} \} . \tag{3.2.73}$$

3.3 THE GENERAL FEEDBACK SYSTEM

In this section we consider the general linear, time-invariant feedback system $\Sigma(\hat{P}, \hat{C})$ shown in Figure 3.9, where $\hat{P} : \begin{bmatrix} v \\ e \end{bmatrix} \mapsto \begin{bmatrix} z \\ y \end{bmatrix}$ represents the $(\eta_o + n_o) \times (\eta_i + n_i)$ plant and $\hat{C} : \begin{bmatrix} v' \\ e' \end{bmatrix} \mapsto \begin{bmatrix} z' \\ y' \end{bmatrix}$ represents the $(\eta_o' + n_i) \times (\eta_i' + n_o)$ compensator. The externally applied inputs are denoted by $\hat{u} := \begin{bmatrix} v \\ u \\ v' \\ u' \end{bmatrix}$, the plant and the compensator outputs are denoted by $\hat{y} := \begin{bmatrix} z \\ y \\ z' \\ y' \end{bmatrix}$; the closed-loop input-output map of $\Sigma(\hat{P}, \hat{C})$ is denoted by $H_{\hat{y}\hat{u}} : \hat{u} \mapsto \hat{y}$.

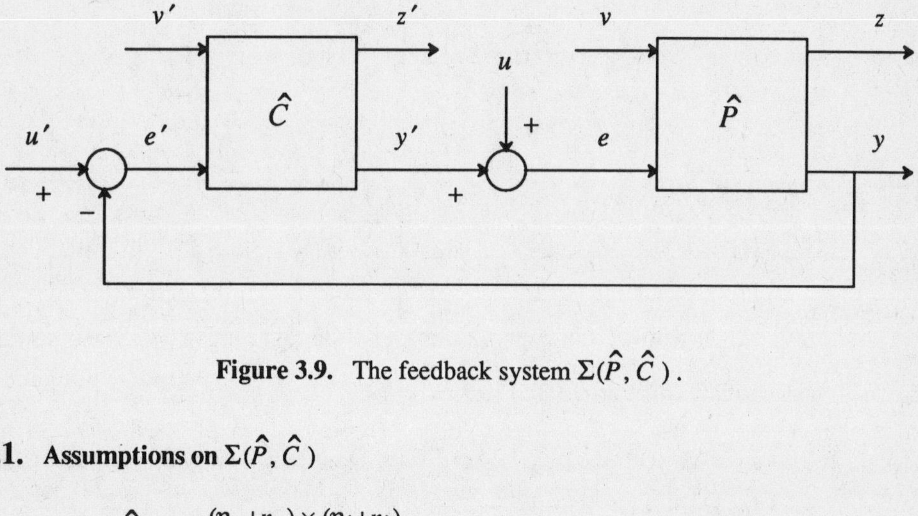

Figure 3.9. The feedback system $\Sigma(\hat{P}, \hat{C})$.

3.3.1. Assumptions on $\Sigma(\hat{P}, \hat{C})$

(i) The plant $\hat{P} \in \mathbf{G}^{(\eta_o+n_o)\times(\eta_i+n_i)}$ and is partitioned as in equation (3.3.1):

$$\hat{P} = \begin{bmatrix} P_{11} & P_{12} \\ P_{21} & P \end{bmatrix} \in \mathbf{G}^{(\eta_o+n_o)\times(\eta_i+n_i)}, \quad \text{where } P \in \mathbf{G}^{n_o\times n_i}. \tag{3.3.1}$$

(ii) The compensator $\hat{C} \in \mathbf{G}^{(\eta_o'+n_i)\times(\eta_i'+n_o)}$ and is partitioned as in equation (3.3.2):

$$\hat{C} = \begin{bmatrix} C_{11} & C_{12} \\ C_{21} & C \end{bmatrix} \in \mathbf{G}^{(\eta_o'+n_i)\times(\eta_i'+n_o)}, \quad \text{where } C \in \mathbf{G}^{n_i\times n_o}. \tag{3.3.2}$$

(iii) The system $\Sigma(\hat{P}, \hat{C})$ is well-posed; equivalently, the closed-loop input-output map $H_{\hat{y}\hat{u}} : \hat{u} \mapsto \hat{y}$ is in $\mathfrak{m}(\mathbf{G})$. □

By Lemma 2.3.5, the plant $\hat{P}$ has an r.c.f.r. $(N_{\hat{P}}, D_{\hat{P}})$ and an l.c.f.r. $(\tilde{D}_{\hat{P}}, \tilde{N}_{\hat{P}})$ which satisfy equations (3.3.3)-(3.3.4) below:

$$(N_{\hat{P}}, D_{\hat{P}}) =: \left(\begin{bmatrix} N_{11} & N_{12} \\ N_{21} & N_p \end{bmatrix}, \begin{bmatrix} D_{11} & 0 \\ D_{21} & D_p \end{bmatrix} \right), \tag{3.3.3}$$

$$(\tilde{D}_{\hat{P}}, \tilde{N}_{\hat{P}}) =: \left(\begin{bmatrix} \tilde{D}_{11} & \tilde{D}_{12} \\ 0 & \tilde{D}_p \end{bmatrix}, \begin{bmatrix} \tilde{N}_{11} & \tilde{N}_{12} \\ \tilde{N}_{21} & \tilde{N}_p \end{bmatrix} \right), \tag{3.3.4}$$

where

$$(N_p, D_p) \text{ is an r.f.r. of } P \text{ and } (\tilde{D}_p, \tilde{N}_p) \text{ is an l.f.r. of } P.$$

By Lemma 2.3.5 applied to $\hat{C}$, the compensator $\hat{C}$ has an l.c.f.r. $(\tilde{D}_{\hat{C}}, \tilde{N}_{\hat{C}})$ and an r.c.f.r. $(N_{\hat{C}}, D_{\hat{C}})$ which satisfy equations (3.3.5)-(3.3.6) below:

$$(\tilde{D}_{\hat{C}}, \tilde{N}_{\hat{C}}) =: \left(\begin{bmatrix} \tilde{D}'_{11} & \tilde{D}'_{12} \\ 0 & \tilde{D}_c \end{bmatrix}, \begin{bmatrix} \tilde{N}'_{11} & \tilde{N}'_{12} \\ \tilde{N}'_{21} & \tilde{N}_c \end{bmatrix} \right), \tag{3.3.5}$$

$$(N_{\hat{C}}, D_{\hat{C}}) =: \left(\begin{bmatrix} N'_{11} & N'_{12} \\ N'_{21} & N_c \end{bmatrix}, \begin{bmatrix} D'_{11} & 0 \\ D'_{21} & D_c \end{bmatrix} \right), \tag{3.3.6}$$

where

$$(\tilde{D}_c, \tilde{N}_c) \text{ is an l.f.r. of } C \quad \text{and } (N_c, D_c) \text{ is an r.f.r. of } C .$$

By Lemma 2.3.4, any *other* r.c.f.r. of $\hat{P}$ is given by $(N_{\hat{p}} R, D_{\hat{p}} R)$, where $(N_{\hat{p}}, D_{\hat{p}})$ is the r.c.f.r in equation (3.3.3) and $R \in \mathbf{m}(\mathrm{H})$ is H–unimodular. Similarly, any other l.c.f.r. of $\hat{P}$ is given by $(L\tilde{D}_{\hat{p}}, L\tilde{N}_{\hat{p}})$, where $(\tilde{D}_{\hat{p}}, \tilde{N}_{\hat{p}})$ is the l.c.f.r. in equation (3.3.4) and $L \in \mathbf{m}(\mathrm{H})$ is H–unimodular. Note that the pair (N_p, D_p) in equation (3.3.3) is *not* necessarily r.c. and the pair $(\tilde{D}_p, \tilde{N}_p)$ in equation (3.3.4) is *not* necessarily l.c. Similar comments apply to coprime-fraction representations of $\hat{C}$.

3.3.2. Closed-loop input-output maps of $\Sigma(\hat{P}, \hat{C})$

Let Assumptions 3.3.1 hold; using (3.3.1)-(3.3.2), the system $\Sigma(\hat{P}, \hat{C})$ in Figure 3.9 is described by:

$$\begin{bmatrix} I_{n_o} & 0 & 0 & -P_{12} \\ 0 & I_{n_o} & 0 & -P \\ 0 & C_{12} & I_{n_o'} & 0 \\ 0 & C & 0 & I_{n_i} \end{bmatrix} \begin{bmatrix} z \\ y \\ z' \\ y' \end{bmatrix} = \begin{bmatrix} P_{11} & P_{12} & 0 & 0 \\ P_{21} & P & 0 & 0 \\ 0 & 0 & C_{11} & C_{12} \\ 0 & 0 & C_{21} & C \end{bmatrix} \begin{bmatrix} v \\ u \\ v' \\ u' \end{bmatrix} . \tag{3.3.7}$$

From equation (3.3.7), it is easy to see that $H_{\hat{y}\hat{u}} : \hat{u} \mapsto \hat{y} \in \mathbf{m}(\mathrm{G})$ if and only if condition (3.2.4) of Section 3.2 holds; (3.2.4) is equivalent to $(I_{n_i} + CP)^{-1} \in \mathbf{m}(\mathrm{G})$ and to $(I_{n_o} + PC)^{-1} \in \mathbf{m}(\mathrm{G})$; consequently, the system $\Sigma(\hat{P}, \hat{C})$ is well-posed if and only if the unity-feedback system $\mathrm{S}(P, C)$ of Section 3.2 is well-posed. By Assumption 3.3.1, since the system $\Sigma(\hat{P}, \hat{C})$ is well-posed, the matrix

$$T := (I_{n_i} + CP)^{-1}$$

is in $\mathbf{m}(\mathrm{G})$. The closed-loop I/O map $H_{\hat{y}\hat{u}}$ is given in terms of T in equation (3.3.8) below; note that it is also possible to write $H_{\hat{y}\hat{u}}$ in terms of $(I_{n_o} + PC)^{-1} \in \mathbf{m}(\mathrm{G})$, where $(I_{n_o} + PC)^{-1} = I_{n_o} - P(I_{n_i} + CP)^{-1}C = I_{n_o} - PTC$:

$$H_{\hat{y}\hat{u}} = \begin{bmatrix} P_{11}-P_{12}T\,C\,P_{21} & P_{12}T & P_{12}T\,C_{21} & P_{12}T\,C \\ (I_{n_o}-P\,T\,C)P_{21} & P\,T & P\,T\,C_{21} & P\,T\,C \\ -C_{12}(I_{n_o}-P\,T\,C)P_{21} & -C_{12}P\,T & C_{11}-C_{12}P\,T\,C_{21} & C_{12}(I_{n_o}-P\,T\,C) \\ -T\,C\,P_{21} & T-I_{n_i} & T\,C_{21} & T\,C \end{bmatrix} \tag{3.3.8}$$

Definition 3.3.3. (**H–stability of** $\Sigma(\hat{P},\hat{C})$)

The system $\Sigma(\hat{P},\hat{C})$ is said to be H–*stable* iff $H_{\hat{y}\hat{u}} \in \mathrm{m}(\mathrm{H})$.

3.3.4. Analysis (Description of $\Sigma(\hat{P},\hat{C})$ using coprime factorizations)

We analyze the system $\Sigma(\hat{P},\hat{C})$ using coprime factorizations of the plant $\hat{P}$ and the compensator $\hat{C}$; we only consider the case where a right-coprime factorization of $\hat{P}$ and a left-coprime factorization of $\hat{C}$ is used.

Let Assumptions 3.3.1 hold. Let $\hat{P} = N_{\hat{p}} D_{\hat{p}}^{-1}$, where ($N_{\hat{p}}, D_{\hat{p}}$) is the r.c.f.r. of $\hat{P}$ given in equation (3.3.3) and let $\hat{C} = \tilde{D}_{\hat{c}}^{-1} \tilde{N}_{\hat{c}}$, where ($\tilde{D}_{\hat{c}}, \tilde{N}_{\hat{c}}$) is the l.c.f.r. of $\hat{C}$ given in equation (3.3.5); any other r.c.f.r. of $\hat{P}$ is given by ($N_{\hat{p}} R$, $D_{\hat{p}} R$) and any other l.c.f.r. of $\hat{C}$ is given by ($\tilde{L}\,\tilde{D}_{\hat{c}}$, $\tilde{L}\,\tilde{N}_{\hat{c}}$), where $R \in \mathrm{m}(\mathrm{H})$, $\tilde{L} \in \mathrm{m}(\mathrm{H})$ are H–unimodular matrices.

The system $\Sigma(\hat{P},\hat{C})$ is redrawn in Figure 3.10; note that $D_{\hat{p}}\,\hat{\xi}_p = \begin{bmatrix} v \\ e \end{bmatrix}$, $N_{\hat{p}}\,\hat{\xi}_p = \begin{bmatrix} z \\ y \end{bmatrix}$, where $\hat{\xi}_p$ denotes the pseudo-state of $\hat{P}$.

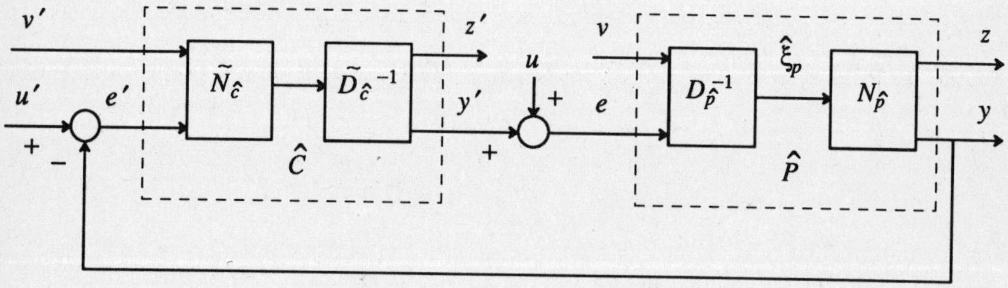

Figure 3.10. The system $\Sigma(\hat{P},\hat{C})$ with $\hat{P} = N_{\hat{p}} D_{\hat{p}}^{-1}$ and $\hat{C} = \tilde{D}_{\hat{c}}^{-1} \tilde{N}_{\hat{c}}$.

The system $\Sigma(\hat{P}, \hat{C})$ is then described by equations (3.3.9)-(3.3.10) for any r.c.f.r. of $\hat{P}$ and any l.c.f.r. of $\hat{C}$; note that the matrix $R \in \mathrm{m}(\mathrm{H})$ appearing in these equations is H–unimodular and that the H–unimodular matrix $\tilde{L} \in \mathrm{m}(\mathrm{H})$ in $(\tilde{L}\tilde{D}_{\hat{c}}, \tilde{L}\tilde{N}_{\hat{c}})$ cancels out:

$$\begin{bmatrix} D_{\hat{p}} R & \vdots & \begin{bmatrix} 0 & 0 \\ 0 & -I_{n_i} \end{bmatrix} \\ \cdots & & \cdots \\ \tilde{N}_{\hat{c}} \begin{bmatrix} 0 & 0 \\ 0 & I_{n_o} \end{bmatrix} N_{\hat{p}} R & \vdots & \tilde{D}_{\hat{c}} \end{bmatrix} \begin{bmatrix} \xi_p \\ \cdots \\ z' \\ y' \end{bmatrix} = \begin{bmatrix} I_{\eta_i + n_i} & \vdots & 0 \\ \cdots & & \cdots \\ 0 & \vdots & \tilde{N}_{\hat{c}} \end{bmatrix} \begin{bmatrix} v \\ u \\ \cdots \\ v' \\ u' \end{bmatrix}, \tag{3.3.9}$$

$$\begin{bmatrix} N_{\hat{p}} R & \vdots & 0 \\ \cdots & & \cdots \\ 0 & \vdots & I_{\eta_o' + n_i} \end{bmatrix} \begin{bmatrix} \xi_p \\ \cdots \\ z' \\ y' \end{bmatrix} = \begin{bmatrix} z \\ y \\ \cdots \\ z' \\ y' \end{bmatrix}. \tag{3.3.10}$$

Equations (3.3.9)-(3.3.10) are of the form

$$\hat{D}_H \hat{\xi} = \hat{N}_L \hat{u}$$

$$\hat{N}_R \hat{\xi} = \hat{y} .$$

By Lemma 2.3.2, performing elementary row operations over $\mathrm{m}(\mathrm{H})$ on the matrix $\begin{bmatrix} \hat{D}_H \\ \hat{N}_R \end{bmatrix}$ and elementary column operations over $\mathrm{m}(\mathrm{H})$ on the matrix $\begin{bmatrix} \hat{N}_L & \hat{D}_H \end{bmatrix}$, we conclude that $(\hat{N}_R, \hat{D}_H, \hat{N}_L)$ is a b.c. triple. Since $\hat{D}_H \in \mathrm{m}(\mathrm{H})$, it follows from Comment 2.4.7 (i) that

$$H_{\hat{y}\hat{u}} = \hat{N}_R \hat{D}_H^{-1} \hat{N}_L \in \mathrm{m}(\mathrm{G}) \tag{3.3.11}$$

if and only if $\det \hat{D}_H \in \mathrm{I}$ (equivalently, the system $\Sigma(\hat{P}, \hat{C})$ is well-posed). Since Assumption 3.3.1 (iii) holds, condition (3.3.11) is satisfied and hence, $\det \hat{D}_H \in \mathrm{I}$. Consequently, $(\hat{N}_R, \hat{D}_H, \hat{N}_L)$ is a b.c.f.r. of $H_{\hat{y}\hat{u}}$ and hence, $\det \hat{D}_H$ is a characteristic determinant of $H_{\hat{y}\hat{u}}$. Using (3.3.3) and (3.3.5) in (3.3.9), we can rewrite $\hat{D}_H$ as:

$$\hat{D}_H = \begin{bmatrix} \begin{bmatrix} D_{11} & 0 \\ D_{21} & D_p \end{bmatrix} & \vdots & \begin{bmatrix} 0 & 0 \\ 0 & -I_{n_i} \end{bmatrix} \\ \begin{bmatrix} \tilde{N}'_{12} N_{21} & \tilde{N}'_{12} N_p \\ \tilde{N}_c N_{21} & \tilde{N}_c N_p \end{bmatrix} & \vdots & \begin{bmatrix} \tilde{D}'_{11} & \tilde{D}'_{12} \\ 0 & \tilde{D}_c \end{bmatrix} \end{bmatrix} \begin{bmatrix} R & \vdots & 0 \\ \cdots & & \cdots \\ 0 & \vdots & I_{\eta_o' + n_i} \end{bmatrix} . \tag{3.3.12}$$

Theorem 3.3.5. (**H–stability of $\Sigma(\hat{P}, \hat{C})$**)

Let Assumptions 3.3.1 (i) and (ii) hold. Let $(N_{\hat{p}}, D_{\hat{p}})$ be the r.c.f.r. and $(\tilde{D}_{\hat{p}}, \tilde{N}_{\hat{p}})$ be the l.c.f.r. of $\hat{P}$ given in equations (3.3.3)-(3.3.4); let $(\tilde{D}_{\hat{c}}, \tilde{N}_{\hat{c}})$ be the l.c.f.r. and $(N_{\hat{c}}, D_{\hat{c}})$ be the r.c.f.r. of $\hat{C}$ given in equations (3.3.5)-(3.3.6). Under these assumptions, the following four statements are equivalent:

(i) $\Sigma(\hat{P}, \hat{C})$ is H–*stable* ;

(ii) $\hat{D}_H$ is H–unimodular ; (3.3.13)

(iii) D_{11} is H–unimodular , and (3.3.14)

$\tilde{D}'_{11}$ is H–unimodular , and (3.3.15)

$\left[\tilde{D}_c D_p + \tilde{N}_c N_p \right]$ is H–unimodular ; (3.3.16)

(iv) $\tilde{D}_{11}$ is H–unimodular , and (3.3.17)

D'_{11} is H–unimodular , and (3.3.18)

$\left[\tilde{D}_p D_c + \tilde{N}_p N_c \right]$ is H–unimodular . (3.3.19)

□

Note that each of statements (i) through (iv) imply that the system $\Sigma(\hat{P}, \hat{C})$ is well-posed; consequently, we do not need a well-posedness assumption in Theorem 3.3.5.

Proof

Statement (i) is equivalent to statement (ii):

The system $\Sigma(\hat{P}, \hat{C})$ is described by equations (3.3.9)-(3.3.10), where by Lemma 2.3.4, $(N_{\hat{P}}R, D_{\hat{P}}R)$ is any arbitrary r.c.f.r. of $\hat{P}$ and $(\tilde{L}\tilde{D}_{\hat{C}}, \tilde{L}\tilde{N}_{\hat{C}})$ is any arbitrary l.c.f.r. of $\hat{C}$ where $R \in \mathfrak{m}(\mathrm{H}), \tilde{L} \in \mathfrak{m}(\mathrm{H})$ are H–unimodular matrices. If $\Sigma(\hat{P}, \hat{C})$ is H–stable, then $H_{\hat{y}\hat{u}} \in \mathfrak{m}(\mathrm{H})$ and hence, condition (3.3.11) holds; equivalently, $\det\hat{D}_H \in \mathrm{I}$; therefore statements (i) and (ii) both imply that $(\hat{N}_R, \hat{D}_H, \hat{N}_L)$ is a b.c.f.r. of $H_{\hat{y}\hat{u}}$. By Lemma 2.4.6, $H_{\hat{y}\hat{u}} \in \mathfrak{m}(\mathrm{H})$ implies that $\hat{D}_H^{-1} \in \mathfrak{m}(\mathrm{H})$. Conversely, if condition (3.3.13) holds, then $\hat{D}_H^{-1} \in \mathfrak{m}(\mathrm{H})$ and hence, $H_{\hat{y}\hat{u}} = \hat{N}_R\, \hat{D}_H^{-1}\, \hat{N}_L \in \mathfrak{m}(\mathrm{H})$.

Statement (ii) is equivalent to statement (iii):

From equation (3.3.12) we calculate $\det\hat{D}_H$ using elementary operations over $\mathfrak{m}(\mathrm{H})$:

$$\det\hat{D}_H = \det D_{11} \det\tilde{D}'_{11} \det(\tilde{D}_c D_p + \tilde{N}_c N_p) \det R \,. \tag{3.3.20}$$

By Lemma 2.3.3 (i), $\det\hat{D}_H \in \mathrm{J}$ if and only if each of the factors in equations (3.3.20) is in J; equivalently, since $R \in \mathfrak{m}(\mathrm{H})$ is H–unimodular, condition (3.3.13) holds if and only if all of (3.3.14), (3.3.15), (3.3.16) hold.

Statement (iii) is equivalent to statement (iv):

By Lemma 2.4.4 applied to $\hat{P}$ and $\hat{C}$, from equations (3.3.3)-(3.3.6) we obtain

$$\det D_{\hat{P}} \sim \det\tilde{D}_{\hat{P}} \quad (\text{equivalently, } \det D_{11} \det D_p \sim \det\tilde{D}_{11} \det\tilde{D}_p) \tag{3.3.21}$$

and

$$\det\tilde{D}_{\hat{C}} \sim \det D_{\hat{C}} \quad (\text{equivalently, } \det\tilde{D}'_{11} \det\tilde{D}_c \sim \det D'_{11} \det D_c)\,. \tag{3.3.22}$$

Since $\det(I_{n_i} + C\,P) = \det(I_{n_o} + P\,C)$, we can rewrite $\det\hat{D}_H$ from (3.3.20) as

$$\det\hat{D}_H = \det D_{11} \det\tilde{D}'_{11} \det\tilde{D}_c \det(I_{n_i} + C\,P) \det D_p \det R$$

$$= (\det D_{11} \det D_p)(\det\tilde{D}'_{11} \det\tilde{D}_c) \det(I_{n_o} + P\,C) \det R \,. \tag{3.3.23}$$

Since $R \in \mathfrak{m}(\mathrm{H})$ is H–unimodular, using equations (3.3.21)-(3.3.22) in (3.3.23) we obtain

$$\det\hat{D}_H \sim (\det\tilde{D}_{11} \det\tilde{D}_p)(\det D'_{11} \det D_c) \det(I_{n_o} + P\, C)\,. \tag{3.3.24}$$

Using $P = \tilde{D}_p^{-1} \tilde{N}_p$ and $C = N_c\, D_c^{-1}$ in (3.3.24) we obtain

$$\det\hat{D}_H \sim \det\tilde{D}_{11} \det D'_{11} \det(\tilde{D}_p\, D_c + \tilde{N}_p\, N_c)\,. \tag{3.3.25}$$

By Lemma 2.3.3 (i), $\det\hat{D}_H \in \mathbf{J}$ if and only if each of the three factors in (3.3.25) is in $\mathbf{J}$; equivalently, condition (3.3.13) holds if and only if all of (3.3.17), (3.3.18), (3.3.19) hold. □

Comment 3.3.6

By Theorem 3.3.5, if $\Sigma(\hat{P}, \hat{C})$ is H–stable, then condition (3.3.14) implies that $\det D_{11} = \det D_{\hat{p}} (\det D_p)^{-1} \in \mathbf{J}$; equivalently, $\det D_{\hat{p}} \sim \det D_p$. Similarly, condition (3.3.17) implies that $\det D_{\hat{p}} \sim \det\tilde{D}_p$. Condition (3.3.16) implies that (N_p, D_p) is an r.c. pair and condition (3.3.19) implies that $(\tilde{D}_p, \tilde{N}_p)$ is an l.c. pair. Therefore, if $\Sigma(\hat{P}, \hat{C})$ is H–stable, then $R := \begin{bmatrix} D_{11}^{-1} & 0 \\ 0 & I_{n_i} \end{bmatrix} \in \mathfrak{m}(\mathrm{H})$ is H–unimodular and $L := \begin{bmatrix} \tilde{D}_{11}^{-1} & 0 \\ 0 & I_{n_o} \end{bmatrix} \in \mathfrak{m}(\mathrm{H})$ is H–unimodular; hence, there exists an r.c.f.r. $(N_{\hat{p}} R, D_{\hat{p}} R)$ of $\hat{P}$ given by

$$(N_{\hat{p}} R, D_{\hat{p}} R) = \left(\begin{bmatrix} N_{11} & N_{12} \\ N_{21} & N_p \end{bmatrix}, \begin{bmatrix} I_{n_i} & 0 \\ D_{21} & D_p \end{bmatrix} \right), \tag{3.3.26}$$

(and (N_p, D_p) is a right-*coprime*-fraction representation of P) and there exists an l.c.f.r. $(L\tilde{D}_{\hat{p}}, L\tilde{N}_{\hat{p}})$ of $\hat{P}$ given by

$$(L\tilde{D}_{\hat{p}}, L\tilde{N}_{\hat{p}}) = \left(\begin{bmatrix} I_{n_o} & \tilde{D}_{12} \\ 0 & \tilde{D}_p \end{bmatrix}, \begin{bmatrix} \tilde{N}_{11} & \tilde{N}_{12} \\ \tilde{N}_{21} & \tilde{N}_p \end{bmatrix} \right), \tag{3.3.27}$$

(and $(\tilde{D}_p, \tilde{N}_p)$ is a left-*coprime*-fraction representation of P). Similar necessary conditions apply to the compensator $\hat{C}$: If $\Sigma(\hat{P}, \hat{C})$ is H–stable, then $\tilde{L} := \begin{bmatrix} \tilde{D}'^{-1}_{11} & 0 \\ 0 & I_{n_i} \end{bmatrix} \in \mathfrak{m}(\mathrm{H})$ is H–unimodular and $\tilde{R} := \begin{bmatrix} D'^{-1}_{11} & 0 \\ 0 & I_{n_o} \end{bmatrix} \in \mathfrak{m}(\mathrm{H})$ is H–unimodular; hence, there exists an l.c.f.r. $(\tilde{L}\tilde{D}_{\hat{C}}, \tilde{L}\tilde{N}_{\hat{C}})$ of $\hat{C}$ given by

$$(\tilde{L}\tilde{D}_{\hat{C}}, \tilde{L}\tilde{N}_{\hat{C}}) = \left(\begin{bmatrix} I_{\eta_o'} & \tilde{D}'_{12} \\ 0 & \tilde{D}_c \end{bmatrix}, \begin{bmatrix} \tilde{N}'_{11} & \tilde{N}'_{12} \\ \tilde{N}'_{21} & \tilde{N}_c \end{bmatrix} \right), \qquad (3.3.28)$$

(and $(\tilde{D}_c, \tilde{N}_c)$ is a left-*coprime*-fraction representation of C) and there exists an r.c.f.r. $(N_{\hat{C}}\tilde{R}, D_{\hat{C}}\tilde{R})$ of $\hat{C}$ given by

$$(N_{\hat{C}}\tilde{R}, D_{\hat{C}}\tilde{R}) = \left(\begin{bmatrix} N'_{11} & N'_{12} \\ N'_{21} & N_c \end{bmatrix}, \begin{bmatrix} I_{\eta_i'} & 0 \\ D'_{21} & D_c \end{bmatrix} \right), \qquad (3.3.29)$$

(and (N_c, D_c) is a right-*coprime*-fraction representation of C). □

Definition 3.3.7. (H–stabilizing compensator $\hat{C}$)

(i) $\hat{C}$ is called an H–*stabilizing compensator for* $\hat{P}$ (abbreviated as: $\hat{C}$ H–*stabilizes* $\hat{P}$) iff $\hat{C} \in \mathbf{G}^{(\eta_o'+n_i)\times(\eta_i'+n_o)}$ *and* the system $\Sigma(\hat{P}, \hat{C})$ is H–stable.

(ii) The set

$$\hat{S}(\hat{P}) := \{ \hat{C} \mid \hat{C} \text{ H–}\textit{stabilizes } \hat{P} \}$$

is called the *set of all* H–*stabilizing compensators for* $\hat{P}$ in the system $\Sigma(\hat{P}, \hat{C})$.

□

Definition 3.3.8. (Σ–admissible plant $\hat{P}$)

(i) $\hat{P} \in \mathfrak{m}(\mathbf{G})$ is called Σ–*admissible* iff there exists a compensator $\hat{C}$ that H–stabilizes $\hat{P}$.

(ii) The class of all Σ–admissible plants is the set of all plants $\hat{P} \in \mathbf{G}^{(\eta_o+n_o)\times(\eta_i+n_i)}$ for which there exists a compensator $\hat{C} \in \mathbf{G}^{(\eta_o'+n_i)\times(\eta_i'+n_o)}$ such that the system $\Sigma(\hat{P}, \hat{C})$ is H–stable. □

Theorem 3.3.9. (Necessary and sufficient conditions for Σ–admissibility)

Let $\hat{P} \in \mathrm{G}^{(\eta_o+n_o)\times(\eta_i+n_i)}$ be as in Assumption 3.3.1 (i); then the following three conditions are equivalent:

(i) $\hat{P}$ is Σ–admissible ;

(ii) $$\left(\begin{bmatrix} \hat{N}_{11} & N_{12} \\ \tilde{V}_p \tilde{N}_{21} & N_p \end{bmatrix}, \begin{bmatrix} I_{\eta_i} & 0 \\ -\tilde{U}_p \tilde{N}_{21} & D_p \end{bmatrix} \right) \text{ is an r.c.f.r. of } \hat{P} \tag{3.3.30}$$

and

$$\left(\begin{bmatrix} I_{\eta_o} & -N_{12} U_p \\ 0 & \tilde{D}_p \end{bmatrix}, \begin{bmatrix} \hat{N}_{11} & N_{12} V_p \\ \tilde{N}_{21} & \tilde{N}_p \end{bmatrix} \right) \text{ is an l.c.f.r. of } \hat{P} , \tag{3.3.31}$$

where (N_p , D_p) is an r.c.f.r. and $(\tilde{D}_p , \tilde{N}_p)$ is an l.c.f.r. of $P \in \mathrm{G}^{n_o \times n_i}$; the matrices U_p , V_p , $\tilde{U}_p$, $\tilde{V}_p$ satisfy the generalized Bezout identity (2.3.12); $\hat{N}_{11} \in \mathrm{H}^{\eta_o \times \eta_i}$, $N_{12} \in \mathrm{H}^{\eta_o \times n_i}$, $\tilde{N}_{21} \in \mathrm{H}^{n_o \times \eta_i}$ are arbitrary H–stable matrices;

(iii) $$P_{11} - P_{12} D_p U_p P_{21} \in \mathfrak{m}(\mathrm{H}) \tag{3.3.32}$$

and $$P_{12} D_p \in \mathfrak{m}(\mathrm{H}) \tag{3.3.33}$$

and $$\tilde{D}_p P_{21} \in \mathfrak{m}(\mathrm{H}) , \tag{3.3.34}$$

where (N_p , D_p) is an r.c.f.r. and $(\tilde{D}_p , \tilde{N}_p)$ is an l.c.f.r. of $P \in \mathrm{G}^{n_o \times n_i}$; the matrices U_p , V_p , $\tilde{U}_p$, $\tilde{V}_p$ satisfy the generalized Bezout identity (2.3.12).

Proof

Statement (i) implies statement (ii):

Suppose that $\hat{P}$ is Σ–admissible; then by Definition 3.3.8, there exists a $\hat{C} \in \mathfrak{m}(\mathrm{G})$ such that $\Sigma(\hat{P}, \hat{C})$ is H–stable. Following Comment 3.3.6, $\hat{P}$ has an r.c.f.r. $(N_{\hat{p}} R , D_{\hat{p}} R)$, given by (3.3.26) and an l.c.f.r. $(L \tilde{D}_{\hat{p}} , L \tilde{N}_{\hat{p}})$, given by (3.3.27), where (N_p , D_p) is an r.c.f.r. and

$(\tilde{D}_p, \tilde{N}_p)$ is an l.c.f.r. of P. Now since $\hat{P} = N_{\hat{P}} D_{\hat{P}}^{-1} = \tilde{D}_{\hat{P}}^{-1} \tilde{N}_{\hat{P}}$ implies that $\tilde{D}_{\hat{P}} N_{\hat{P}} = \tilde{N}_{\hat{P}} D_{\hat{P}}$ (and therefore, $L \tilde{D}_{\hat{P}} N_{\hat{P}} R = L \tilde{N}_{\hat{P}} D_{\hat{P}} R$), from (3.3.26)-(3.3.27) we obtain the following equations:

$$\tilde{D}_p N_p = \tilde{N}_p D_p \ , \tag{3.3.35}$$

$$\tilde{N}_{12} D_p + (-\tilde{D}_{12}) N_p = N_{12} \ , \tag{3.3.36}$$

$$\tilde{N}_p (-D_{21}) + \tilde{D}_p N_{21} = \tilde{N}_{21} \ , \tag{3.3.37}$$

$$\tilde{N}_{12} D_{21} - \tilde{D}_{12} N_{21} = N_{11} - \tilde{N}_{11} \ . \tag{3.3.38}$$

Since (N_p, D_p) is an r.c.f.r. and $(\tilde{D}_p, \tilde{N}_p)$ is an l.c.f.r. of P, equation (3.3.35) implies that $((N_p, D_p), (\tilde{D}_p, \tilde{N}_p))$ is a doubly-coprime pair. Let V_p, U_p, $\tilde{V}_p$, $\tilde{U}_p$ be as in the generalized Bezout identity (2.3.12). By Lemma 2.5.1 (i), $(\tilde{N}_{12}, \tilde{D}_{12})$ is a solution of equation (3.3.36) over $\mathfrak{m}(\mathrm{H})$ if and only if

$$\begin{bmatrix} \tilde{N}_{12} & -\tilde{D}_{12} \end{bmatrix} = \begin{bmatrix} N_{12} & \hat{Q} \end{bmatrix} \begin{bmatrix} V_p & U_p \\ -\tilde{N}_p & \tilde{D}_p \end{bmatrix} , \tag{3.3.39}$$

for some $\hat{Q} \in \mathfrak{m}(\mathrm{H})$; similarly, by Lemma 2.5.1 (ii), (D_{21}, N_{21}) is a solution of equation (3.3.37) over $\mathfrak{m}(\mathrm{H})$ if and only if

$$\begin{bmatrix} D_{21} \\ N_{21} \end{bmatrix} = \begin{bmatrix} D_p & -\tilde{U}_p \\ N_p & \tilde{V}_p \end{bmatrix} \begin{bmatrix} -\tilde{Q} \\ \tilde{N}_{21} \end{bmatrix} , \tag{3.3.40}$$

for some $\tilde{Q} \in \mathfrak{m}(\mathrm{H})$. Substituting for $(\tilde{N}_{12}, \tilde{D}_{12})$ and (D_{21}, N_{21}) from equations (3.3.39)-(3.3.40) into (3.3.38) and using the generalized Bezout identity (2.3.12) we obtain

$$\begin{bmatrix} \tilde{N}_{12} & -\tilde{D}_{12} \end{bmatrix} \begin{bmatrix} D_{21} \\ N_{21} \end{bmatrix} = \begin{bmatrix} N_{12} & \hat{Q} \end{bmatrix} \begin{bmatrix} -\tilde{Q} \\ \tilde{N}_{21} \end{bmatrix} = N_{11} - \tilde{N}_{11} \ . \tag{3.3.41}$$

Using equations (3.3.39)-(3.3.40), the r.c.f.r. $(N_{\hat{P}} R, D_{\hat{P}} R)$ and the l.c.f.r. $(L \tilde{D}_{\hat{P}}, L \tilde{N}_{\hat{P}})$ of $\hat{P}$ become:

$$(N_{\hat{P}} R, D_{\hat{P}} R) = \left(\begin{bmatrix} N_{11} & N_{12} \\ \tilde{V}_p \tilde{N}_{21} - N_p \tilde{Q} & N_p \end{bmatrix}, \begin{bmatrix} I_{\eta_i} & 0 \\ -\tilde{U}_p \tilde{N}_{21} - D_p \tilde{Q} & D_p \end{bmatrix} \right) \quad (3.3.42)$$

$$(L \tilde{D}_{\hat{P}}, L \tilde{N}_{\hat{P}}) = \left(\begin{bmatrix} I_{\eta_o} & -N_{12} U_p - \hat{Q} \tilde{D}_p \\ 0 & \tilde{D}_p \end{bmatrix}, \begin{bmatrix} \tilde{N}_{11} & N_{12} V_p - \hat{Q} \tilde{N}_p \\ \tilde{N}_{21} & \tilde{N}_p \end{bmatrix} \right) \quad (3.3.43)$$

Let $\hat{R} := \begin{bmatrix} I_{\eta_i} & 0 \\ \tilde{Q} & I_{n_i} \end{bmatrix}$ and let $\hat{L} := \begin{bmatrix} I_{\eta_o} & \hat{Q} \\ 0 & I_{n_o} \end{bmatrix}$; the matrices $\hat{R}$, $\hat{L} \in \mathrm{m}(\mathrm{H})$ are H–unimodular for all $\tilde{Q}$, $\hat{Q} \in \mathrm{m}(\mathrm{H})$. Let $\hat{N}_{11} := N_{11} + N_{12} \tilde{Q}$; by equation (3.3.41),

$$\tilde{N}_{11} + \hat{Q} \tilde{N}_{21} = N_{11} + N_{12} \tilde{Q} = \hat{N}_{11} . \quad (3.3.44)$$

Since $\hat{R}$ and $\hat{L}$ are H–unimodular, by Lemma 2.3.4, $(N_{\hat{P}} R \hat{R}, D_{\hat{P}} R \hat{R})$ is also an r.c.f.r. and $(\hat{L} L \tilde{D}_{\hat{P}}, \hat{L} L \tilde{N}_{\hat{P}})$ is also an l.c.f.r. of $\hat{P}$; but these are the r.c.f.r. given in (3.3.30) and the l.c.f.r. given in (3.3.31).

Statement (ii) implies statement (iii):

Suppose that statement (ii) of Theorem 3.3.9 holds. By (3.3.30) we have

$$\hat{P} = \begin{bmatrix} P_{11} & P_{12} \\ P_{21} & P \end{bmatrix} = \begin{bmatrix} \hat{N}_{11} & N_{12} \\ \tilde{V}_p \tilde{N}_{21} & N_p \end{bmatrix} \begin{bmatrix} I_{\eta_i} & 0 \\ D_p^{-1} \tilde{U}_p \tilde{N}_{21} & D_p^{-1} \end{bmatrix}$$

$$= \begin{bmatrix} \hat{N}_{11} + N_{12} D_p^{-1} \tilde{U}_p \tilde{N}_{21} & N_{12} D_p^{-1} \\ \tilde{V}_p \tilde{N}_{21} + N_p D_p^{-1} \tilde{U}_p \tilde{N}_{21} & N_p D_p^{-1} \end{bmatrix} ; \quad (3.3.45)$$

by (3.3.31) we have

$$\hat{P} = \begin{bmatrix} I_{\eta_o} & -N_{12} U_p \tilde{D}_p^{-1} \\ 0 & \tilde{D}_p^{-1} \end{bmatrix} \begin{bmatrix} \hat{N}_{11} & N_{12} V_p \\ \tilde{N}_{21} & \tilde{N}_p \end{bmatrix}$$

$$= \begin{bmatrix} \hat{N}_{11} + N_{12} U_p \tilde{D}_p^{-1} \tilde{N}_{21} & N_{12} V_p + N_{12} U_p \tilde{D}_p^{-1} \tilde{N}_p \\ \tilde{D}_p^{-1} \tilde{N}_{21} & \tilde{D}_p^{-1} \tilde{N}_p \end{bmatrix} . \qquad (3.3.46)$$

Using the expression for P_{11} and P_{21} in (3.3.46) and for P_{12} in (3.3.45), we obtain

$$P_{11} - P_{12} D_p U_p P_{21} = \hat{N}_{11} \in \mathrm{m}(\mathrm{H}) \qquad (3.3.47)$$

$$P_{12} D_p = N_{12} \in \mathrm{m}(\mathrm{H}) \qquad (3.3.48)$$

$$\tilde{D}_p P_{21} = \tilde{N}_{21} \in \mathrm{m}(\mathrm{H}) . \qquad (3.3.49)$$

Statement (iii) follows then from (3.3.47)-(3.3.49).

Statement (iii) implies statement (i):

Suppose that statement (iii) of Theorem 3.3.9 holds. Let $Q \in \mathrm{m}(\mathrm{H})$ be such that $\det(V_p - Q \tilde{N}_p) \in \mathrm{I}$; note that (2.3.17) holds. Choose the compensator $\hat{C}$ as

$$\hat{C} = \begin{bmatrix} C_{11} & C_{12} \\ C_{21} & C \end{bmatrix} = \begin{bmatrix} I_{\eta_o'} & 0 \\ 0 & V_p - Q \tilde{N}_p \end{bmatrix}^{-1} \begin{bmatrix} 0 & 0 \\ 0 & U_p + Q \tilde{D}_p \end{bmatrix}$$

$$= \begin{bmatrix} 0 & 0 \\ 0 & (V_p - Q \tilde{N}_p)^{-1} (U_p + Q \tilde{D}_p) \end{bmatrix} \in \mathrm{m}(\mathrm{G}) . \qquad (3.3.50)$$

With $C = (V_p - Q \tilde{N}_p)^{-1} (U_p + Q \tilde{D}_p) = \tilde{D}_c^{-1} \tilde{N}_c$ as in (3.3.50), using the generalized Bezout identity (2.3.17), $T := (I_{n_i} + C P)^{-1}$ beomes

$$T = D_p (V_p - Q \tilde{N}_p) . \qquad (3.3.51)$$

Using (3.3.51), (2.3.17) and the compensator $\hat{C} \in \mathrm{m}(\mathrm{G})$ given by (3.3.50), we rewrite the I/O map $H_{\hat{y}\hat{u}}$ given by (3.3.8) in equation (3.3.52) below:

$$H_{\hat{y}\hat{u}} = \begin{bmatrix} P_{11} - P_{12} D_p (U_p + Q\tilde{D}_p) P_{21} & P_{12} D_p (V_p - Q\tilde{N}_p) & 0 & P_{12} D_p (U_p + Q\tilde{D}_p) \\ (\tilde{V}_p - N_p Q) \tilde{D}_p P_{21} & N_p (V_p - Q \tilde{N}_p) & 0 & N_p (U_p + Q \tilde{D}_p) \\ 0 & 0 & 0 & 0 \\ (\tilde{U}_p + D_p Q) \tilde{D}_p P_{21} & -(\tilde{U}_p + D_p Q) \tilde{N}_p & 0 & D_p (U_p + Q \tilde{D}_p) \end{bmatrix} . \qquad (3.3.52)$$

Now conditions (3.3.32)-(3.3.34) imply that all of the maps in (3.3.52) are H–stable. Since $H_{\hat{y}\hat{u}} \in \mathrm{m}(\mathrm{H})$ with this choice of $\hat{C} \in \mathrm{m}(\mathrm{G})$, it follows from Definition 3.3.8 (i) that $\hat{P}$ is Σ–admissible. □

We now parametrize the set $\hat{S}(\hat{P})$ of all H–stabilizing compensators for $\hat{P}$.

Theorem 3.3.10. (Parametrization of all H–stabilizing compensators in $\Sigma(\hat{P}, \hat{C})$)

Let Assumptions 3.3.1 (i) and (iii) hold. Let $\hat{P} \in \mathrm{m}(\mathrm{G})$ be Σ–admissible. Let (N_p , D_p) be an r.c.f.r. and $(\tilde{D}_p , \tilde{N}_p)$ be an l.c.f.r. of $P \in \mathrm{H}^{n_o \times n_i}$; let V_p , U_p , $\tilde{V}_p$, $\tilde{U}_p$ be as in the generalized Bezout identity (2.3.12). Under these assumptions, the set $\hat{S}(\hat{P})$ of all H–stabilizing compensators $\hat{C}$ for $\hat{P}$ is given by equation (3.3.53) and equivalently, by equation (3.3.54) below:

$$\hat{S}(\hat{P}) = \{ \hat{C} = \begin{bmatrix} I_{\eta_o'} & -Q_{12} \tilde{N}_p \\ 0 & V_p - Q \tilde{N}_p \end{bmatrix}^{-1} \begin{bmatrix} Q_{11} & Q_{12} \tilde{D}_p \\ Q_{21} & U_p + Q \tilde{D}_p \end{bmatrix} \;\Big|$$

$$Q_{11} \in \mathrm{H}^{\eta_o' \times \eta_i'} , \; Q_{12} \in \mathrm{H}^{\eta_o' \times n_o} , \; Q_{21} \in \mathrm{H}^{n_i \times \eta_i'} , \; Q \in \mathrm{H}^{n_i \times n_o} ,$$

$$\det(V_p - Q \tilde{N}_p) \in \mathrm{I} \; \} , \qquad (3.3.53)$$

$$\hat{S}(\hat{P}) = \{ \hat{C} = \begin{bmatrix} Q_{11} & Q_{12} \\ D_p Q_{21} & \tilde{U}_p + D_p Q \end{bmatrix} \begin{bmatrix} I_{\eta_i'} & 0 \\ -N_p Q_{21} & \tilde{V}_p - N_p Q \end{bmatrix}^{-1} \mid$$

$$Q_{11} \in \mathrm{H}^{\eta_o' \times \eta_i'} , Q_{12} \in \mathrm{H}^{\eta_o' \times n_o} , Q_{21} \in \mathrm{H}^{n_i \times \eta_i'} , Q \in \mathrm{H}^{n_i \times n_o} ,$$

$$\det(\tilde{V}_p - N_p Q) \subset \mathrm{I} \} , \tag{3.3.54}$$

Furthermore, corresponding to each compensator $\hat{C} \in \hat{S}(\hat{P})$, there is a unique Q_{11}, a unique Q_{12}, a unique Q_{21} and a unique $Q \in \mathrm{m}(\mathrm{H})$ in the equivalent parametrizations (3.3.53) and (3.3.54). Equations (3.3.53) and (3.3.54) are bijections from Q_{11}, Q_{12}, Q_{21}, $Q \in \mathrm{m}(\mathrm{H})$ to $\hat{C} \in \hat{S}(\hat{P})$. □

Proof

Following Comment (3.3.6) and by Theorem 3.3.5, if $\hat{C}$ is is an H–stabilizing compensator for $\hat{P}$, then $\hat{C}$ has an l.c.f.r. $(\tilde{L}\tilde{D}_{\hat{c}} , \tilde{L}\tilde{N}_{\hat{c}})$ given by (3.3.28) and an r.c.f.r. $(N_{\hat{c}}\tilde{R} , D_{\hat{c}}\tilde{R})$ given by (3.3.29), where the l.c.f.r. $(\tilde{D}_c , \tilde{N}_c)$ and the r.c.f.r. (N_c , D_c) of C satisfy conditions (3.3.16) and (3.3.19), respectively. By Lemma 2.5.1, the set of all solutions of (3.3.16) for $(\tilde{D}_c , \tilde{N}_c)$ is given by (2.5.7) and the set of all solutions of (3.3.17) for (N_c , D_c) is given by (2.5.8). Note that (3.3.16)-(3.3.17) are equivalent to (3.2.29) by Corollary 3.2.9.

Now since $\hat{C} = \tilde{D}_{\hat{c}}^{-1}\tilde{N}_{\hat{c}} = N_{\hat{c}}D_{\hat{c}}^{-1}$ implies that $\tilde{D}_{\hat{c}} N_{\hat{c}} = \tilde{D}_{\hat{c}} N_{\hat{c}}$, from (3.3.28)-(3.3.29) we obtain the following equations:

$$\tilde{D}_c N_c = \tilde{N}_c D_c \tag{3.3.55}$$

$$\tilde{N}'_{12} D_c + (-\tilde{D}'_{12}) N_c = N'_{12} , \tag{3.3.56}$$

$$\tilde{N}_c (-D'_{21}) + \tilde{D}_c N'_{21} = \tilde{N}'_{21} , \tag{3.3.57}$$

$$\tilde{N}'_{12} D'_{21} - \tilde{D}'_{12} N'_{21} = N'_{11} - \tilde{N}'_{11} . \tag{3.3.58}$$

Equations (3.3.55)-(3.3.58) for $\hat{C}$ are similar to equations (3.3.35)-(3,3.38) for $\hat{P}$. Following similar steps as in the proof of Theorem 3.3.9 and using the generalized Bezout identity (3.2.29), it is easy to show that $(\tilde{N}'_{12}, \tilde{D}'_{12})$ is a solution of (3.3.56) if and only if

$$\begin{bmatrix} \tilde{N}'_{12} & -\tilde{D}'_{12} \end{bmatrix} = \begin{bmatrix} N'_{12} & \hat{Q}' \end{bmatrix} \begin{bmatrix} \tilde{D}_p & \tilde{N}_p \\ -\tilde{N}_c & \tilde{D}_c \end{bmatrix}, \tag{3.3.59}$$

and (D'_{21}, N'_{21}) is a solution of (3.3.57) if and only if

$$\begin{bmatrix} D'_{21} \\ N'_{21} \end{bmatrix} = \begin{bmatrix} D_c & -N_p \\ N_c & D_p \end{bmatrix} \begin{bmatrix} -\tilde{Q}' \\ \tilde{N}'_{21} \end{bmatrix}, \tag{3.3.60}$$

for some $\hat{Q}' \in \mathfrak{m}(\mathrm{H})$ and $\tilde{Q}' \in \mathfrak{m}(\mathrm{H})$. Substituting from (3.3.59)-(3.3.60) into (3.3.58) and using the generalized Bezout identity (3.2.29) we obtain

$$\begin{bmatrix} \tilde{N}'_{12} & \tilde{D}'_{12} \end{bmatrix} \begin{bmatrix} D'_{21} \\ N'_{21} \end{bmatrix} = \begin{bmatrix} N'_{12} & \hat{Q}' \end{bmatrix} \begin{bmatrix} -\tilde{Q}' \\ \tilde{N}'_{21} \end{bmatrix} = N'_{11} - \tilde{N}'_{11} \tag{3.3.61}$$

Let $L' := \begin{bmatrix} I_{n_o'} & \hat{Q}' \\ 0 & I_{n_i} \end{bmatrix}$ and let $R' := \begin{bmatrix} I_{n_i'} & 0 \\ \tilde{Q}' & I_{n_o} \end{bmatrix}$; the matrices L', $R' \in \mathfrak{m}(\mathrm{H})$ are H–unimodular for all $\tilde{Q}', \hat{Q}' \in \mathfrak{m}(\mathrm{H})$. Let $Q_{11} := \tilde{N}'_{11} + \hat{Q}'\tilde{N}'_{21}$; by equation (3.3.61),

$$\tilde{N}'_{11} + \hat{Q}'\tilde{N}'_{21} = N_{11} + N'_{12}\tilde{Q}' = Q_{11} \; . \tag{3.3.62}$$

Now let $Q_{12} := N'_{12}$ and let $Q_{21} := \tilde{N}'_{21}$. Since L' and R' are H–unimodular, by Lemma 2.3.4, $(L'\tilde{L}\tilde{D}_{\hat{c}}, L'\tilde{L}\tilde{N}_{\hat{c}})$ is also an l.c.f.r. of $\hat{C}$ and $(N_{\hat{c}}\tilde{R}R', D_{\hat{c}}\tilde{R}R')$ is also an r.c.f.r. of $\hat{C}$, where

$$(L'\tilde{L}\tilde{D}_{\hat{c}}, L'\tilde{L}\tilde{N}_{\hat{c}}) = (\begin{bmatrix} I_{n_o'} & -Q_{12}\tilde{N}_p \\ 0 & \tilde{D}_c \end{bmatrix}, \begin{bmatrix} Q_{11} & Q_{12}\tilde{D}_p \\ Q_{21} & \tilde{N}_c \end{bmatrix}), \tag{3.3.63}$$

$$(N_{\hat{C}} \tilde{R} R', D_{\hat{C}} \tilde{R} R') = \left(\begin{bmatrix} Q_{11} & Q_{12} \\ D_p Q_{21} & N_c \end{bmatrix}, \begin{bmatrix} I_{\eta_i'} & 0 \\ -N_p Q_{21} & D_c \end{bmatrix} \right), \tag{3.3.64}$$

and

$$(\tilde{D}_c, \tilde{N}_c) = ((V_p - Q \tilde{N}_p), (U_p + Q \tilde{D}_p)), \tag{3.3.65}$$

$$(N_c, D_c) = ((\tilde{U}_p + D_p Q), (\tilde{V}_p - N_p Q)), \tag{3.3.66}$$

and $Q \in \mathrm{m}(\mathrm{H})$ is such that $\det(V_p - Q \tilde{N}_p) \sim \det(\tilde{V}_p - N_p Q) \in \mathrm{I}$. Thus we showed that all H–stabilizing compensators $\hat{C}$ for $\hat{P}$ are as in the equivalent expressions (3.3.53) and (3.3.54).

Now let $\hat{C}_1$ and $\hat{C}_2$ be two H–stabilizing compensators for $\hat{P}$; by (3.3.53)-(3.3.54),

$$\hat{C}_1 = \begin{bmatrix} Q_{11} + Q_{12} \tilde{N}_p \tilde{D}_{c1}^{-1} Q_{21} & Q_{12} (\tilde{D}_p + \tilde{N}_p C_1) \\ \tilde{D}_{c1}^{-1} Q_{21} & C_1 \end{bmatrix}$$

and

$$\hat{C}_2 = \begin{bmatrix} \hat{Q}_{11} + \hat{Q}_{12} \tilde{N}_p \tilde{D}_{c2}^{-1} \hat{Q}_{21} & \hat{Q}_{12} (\tilde{D}_p + \tilde{N}_p C_2) \\ \tilde{D}_{c2}^{-1} \hat{Q}_{21} & C_2 \end{bmatrix},$$

where $C_1 = \tilde{D}_{c1}^{-1} \tilde{N}_{c1} = (V_p - Q_1 \tilde{N}_p)^{-1} (U_p + Q_1 \tilde{D}_p)$, $C_2 = \tilde{D}_{c2}^{-1} \tilde{N}_{c2} = (V_p - Q_2 \tilde{N}_p)^{-1} (U_p + Q_2 \tilde{D}_p)$. From the proof of Theorem 3.2.11, $C_1 = C_2$ if and only if $Q_1 = Q_2$ and $\tilde{D}_{c1} = \tilde{D}_{c2}$. If $C_1 = C_2$, then $D_{c1}^{-1} = (\tilde{D}_p + \tilde{N}_p C_1) = (\tilde{D}_p + \tilde{N}_p C_2) = D_{c2}^{-1}$. Now $\hat{C}_1 = \hat{C}_2$ implies that: $C_{12} = \hat{C}_{12} = Q_{12} D_{c1}^{-1} = \hat{Q}_{12} D_{c2}^{-1}$ and hence, $Q_{12} = \hat{Q}_{12}$; $C_{21} = \hat{C}_{21} = \tilde{D}_{c1}^{-1} Q_{21} = \tilde{D}_{c2}^{-1} \hat{Q}_{21}$ and hence, $Q_{21} = \hat{Q}_{21}$; $C_{11} = \hat{C}_{11} = Q_{11} + Q_{12} \tilde{N}_p \tilde{D}_{c1}^{-1} Q_{21} = \hat{Q}_{11} + \hat{Q}_{12} \tilde{N}_p \tilde{D}_{c2}^{-1} \hat{Q}_{21}$ and hence, $Q_{11} = \hat{Q}_{11}$ since $Q_{12} = \hat{Q}_{12}$ and $Q_{21} = \hat{Q}_{21}$. We conclude that there is a unique set of (matrix-) parameters Q_{11}, Q_{12}, Q_{21}, $Q \in \mathrm{m}(\mathrm{H})$ corresponding to each H–stabilizing compensator $\hat{C} \in \hat{S}(\hat{P})$.

Comment 3.3.11

(i) Suppose that $\hat{P}$ is Σ–admissible; then by Theorem 3.3.9, $\hat{P}$ is given by (3.3.45) and equivalently, by (3.3.46). Now suppose that $\hat{C}$ is an H–stabilizing compensator for $\hat{P}$; then by Theorem 3.3.10, $\hat{C}$ is given by

$$\hat{C} = \begin{bmatrix} I_{\eta_o'} & -Q_{12}\tilde{N}_p \\ 0 & V_p - Q\tilde{N}_p \end{bmatrix}^{-1} \begin{bmatrix} Q_{11} & Q_{12}\tilde{D}_p \\ Q_{21} & U_p + Q\tilde{D}_p \end{bmatrix}, \qquad (3.3.68)$$

where $Q \in \mathrm{m}(\mathrm{H})$ is such that $\det(V_p - Q\tilde{N}_p) \in \mathrm{I}$. Figure 3.11 shows $\Sigma(\hat{P}, \hat{C})$ where, from (3.3.30), $\hat{P} = N_{\hat{p}} D_{\hat{p}}^{-1}$ is given by

$$\hat{P} = N_{\hat{p}} D_{\hat{p}}^{-1} = \begin{bmatrix} \hat{N}_{11} & N_{12} \\ \tilde{V}_p \tilde{N}_{21} & N_p \end{bmatrix} \begin{bmatrix} I_{\eta_i} & 0 \\ D_p^{-1}\tilde{U}_p \tilde{N}_{21} & D_p^{-1} \end{bmatrix} \qquad (3.3.69)$$

and $\hat{C} = \tilde{D}_{\hat{c}}^{-1} \tilde{N}_{\hat{c}}$ is given by (3.3.68). Note that the only maps in $\hat{P}$ and $\hat{C}$ that may not be H–stable are D_p^{-1} and $\tilde{D}_c^{-1} = (V_p - Q\tilde{N}_p)^{-1}$, respectively.

Suppose that the ring H is the ring of proper stable rational functions R_{U} as in Section 2.2. In this case if $\hat{P}$ is Σ–admissible, then every U–pole of P_{11}, P_{12} and of P_{21} is a U–pole of $P = N_p D_p^{-1}$ with *at most* the same McMillan degree. For $\hat{C}$ to be an H–stabilizing compensator for $\hat{P}$, the U–poles of each of C_{11}, C_{12}, C_{21} must be a subset of the U–poles of $C = \tilde{D}_c^{-1}\tilde{N}_c$, where C is chosen so that the feedback-loop is H–stable.

(ii) The class of all H–stabilizing compensators is *parametrized* by four matrices, Q_{11}, Q_{12}, Q_{21}, $Q \in \mathrm{m}(\mathrm{H})$; the matrix Q parametrizes the class of all C that H–stabilizes the loop. We refer to design with the unity-feedback system $\mathrm{S}(P, C)$ as *one-degree-of-freedom* design because only one parameter matrix is available for design (see Section 3.2). In contrast, for the more general system $\Sigma(\hat{P}, \hat{C})$, there are *four-degrees-of-freedom* because $\hat{C}$ has four completely free matrices in H, which can be chosen to meet performance specifications. For example, the parameter Q_{21} can be used to diagonalize the input-output map $H_{zv'} : v' \mapsto z$.

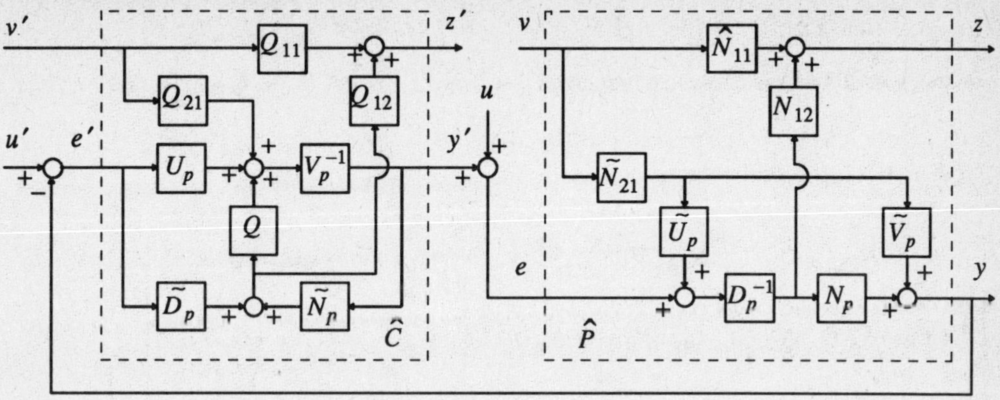

Figure 3.11. The system $\Sigma(\hat{P},\hat{C})$ with a Σ–admissible plant $\hat{P} = N_{\hat{p}} D_{\hat{p}}^{-1}$ and an H–stabilizing compensator $\hat{C}$.

3.3.12. Achievable input-output maps of $\Sigma(\hat{P},\hat{C})$

We can now describe the set of all achievable closed-loop I/O maps

$$H_{\hat{y}\hat{u}} : \begin{bmatrix} v \\ u \\ v' \\ u' \end{bmatrix} \mapsto \begin{bmatrix} z \\ y \\ z' \\ y' \end{bmatrix}$$

of $\Sigma(\hat{P},\hat{C})$ based on the parametrization of all H–stabilizing compensators $\hat{S}(\hat{P})$ given in the equivalent representations (3.3.53) and (3.3.54).

The set

$$\hat{\mathrm{A}}(\hat{P}) := \{ \; H_{\hat{y}\hat{u}} \; | \; \hat{C} \text{ H–stabilizes } \hat{P} \; \}$$

is called *the set of all achievable I/O maps* of the system $\Sigma(\hat{P},\hat{C})$.

By Theorem 3.3.10, $\hat{\mathrm{A}}(\hat{P}) = \{ \; H_{\hat{y}\hat{u}} \; | \; \hat{C} \in \hat{S}(\hat{P}) \; \}$, where $\hat{S}(\hat{P})$ is the set of all H–stabilizing compensators given by (3.3.53) and equivalently, (3.3.54). We obtain the set of all achievable I/O maps from the expression (3.3.8) for $H_{\hat{y}\hat{u}}$ by using the expression for $\hat{P}$ given in (3.3.69) and the expression for $\hat{C}$ given in (3.3.68) and the generalized Bezout identity

(2.3.17). Note that $C = (V_p - Q \tilde{N}_p)^{-1} (U_p + Q \tilde{D}_p) = (\tilde{U}_p + D_p Q) (\tilde{V}_p - N_p Q)^{-1}$ where $Q \in \mathrm{m}(\mathrm{H})$ is chosen so that $\det(V_p - Q \tilde{N}_p) \sim \det(\tilde{V}_p - N_p Q) \in \mathrm{I}$.

$$\hat{\mathrm{A}}(\hat{P}) = \{ H_{\hat{y}\hat{u}} =$$

$$\begin{bmatrix} \hat{N}_{11} - N_{12} Q \tilde{N}_{21} & N_{12} (V_p - Q \tilde{N}_p) & N_{12} Q_{21} & N_{12} (U_p + Q \tilde{D}_p) \\ (\tilde{V}_p - N_p Q) \tilde{N}_{21} & N_p (V_p - Q \tilde{N}_p) & N_p Q_{21} & N_p (U_p + Q \tilde{D}_p) \\ - Q_{12} \tilde{N}_{21} & - Q_{12} \tilde{N}_p & Q_{11} & Q_{12} \tilde{D}_p \\ -(\tilde{U}_p + D_p Q) \tilde{N}_{21} & -(\tilde{U}_p + D_p Q) \tilde{N}_p & D_p Q_{21} & D_p (U_p + Q \tilde{D}_p) \end{bmatrix} \Bigg|$$

$$Q_{11} , Q_{12} , Q_{21} , Q \in \mathrm{m}(\mathrm{H}) , \det(V_p - Q \tilde{N}_p) \sim \det(\tilde{V}_p - N_p Q) \in \mathrm{I} \} . \qquad (3.3.70)$$

The representation (3.3.70) is a parametrization of all closed-loop I/O maps. Each of the I/O maps of $H_{\hat{y}\hat{u}}$ in (3.3.70) is affine in one of the (matrix-) parameters $Q_{11} , Q_{12} , Q_{21} , Q$; these matrices are the design parameters. The system $\Sigma(\hat{P}, \hat{C})$ is called a *four-parameter* design or a *four-degrees-of-freedom* compensation scheme.

If $P_{11} = 0$ and $P_{21} = I_{n_o}$, then the input v can be viewed as an additive disturbance at the plant-output y . If $P_{21} = I_{n_o}$, then $\tilde{D}_p = \tilde{N}_{21}$ by (3.3.46). The disturbance-to-output map $H_{yv} : v \mapsto y$ is given by $(\tilde{V}_p - N_p Q) \tilde{N}_{21} = (\tilde{V}_p - N_p Q) \tilde{D}_p$, which depends on the parameter $Q \in \mathrm{m}(\mathrm{H})$. On the other hand, the external-input to output maps $H_{zv'} = N_{12} Q_{21}$ and $H_{yv'} = N_p Q_{21}$ depend on a *different* parameter Q_{21} .

3.3.13. Decoupling in $\Sigma(\hat{P}, \hat{C})$

Consider the system $\Sigma(\hat{P}, \hat{C})$; let Assumptions 3.3.1 hold; let $\hat{P} \in \mathrm{m}(\mathrm{G})$, partitioned as in equation (3.3.1), be a Σ–admissible plant. Assume that $\eta_i' = \eta_o$; i.e., the number of inputs v' and the number of outputs z are equal. We now consider the problem of decoupling in the system $\Sigma(\hat{P}, \hat{C})$.

Let $(N_{\hat{P}}, D_{\hat{P}})$ be an r.c.f.r. and $(\tilde{D}_{\hat{P}}, \tilde{N}_{\hat{P}})$ be an l.c.f.r. of $\hat{P}$; since $\hat{P}$ is Σ–admissible, we can assume without loss of generality that $(N_{\hat{P}}, D_{\hat{P}})$ is given by (3.3.30) and $(\tilde{D}_{\hat{P}}, \tilde{N}_{\hat{P}})$ is given by (3.3.31).

The system $\Sigma(\hat{P}, \hat{C})$ is said to be *decoupled* iff $\Sigma(\hat{P}, \hat{C})$ is $\mathbf{H}$–stable and the closed-loop map $H_{zv'} : v' \mapsto z$ from the external-input v' to the actual-output z of the plant is *diagonal* and *nonsingular*; a compensator $\hat{C}$ is said to decouple the system $\Sigma(\hat{P}, \hat{C})$ iff $\hat{C} \in \hat{S}(\hat{P})$ and the map $H_{zv'} : v' \mapsto z$ is *diagonal* and *nonsingular*.

By equation (3.3.70), $H_{zv'}$ is an achievable map of $\Sigma(\hat{P}, \hat{C})$ if and only if

$$H_{zv'} = N_{12} Q_{21} \tag{3.3.71}$$

for some $Q_{21} \in \mathbf{H}^{n_i \times \eta_o}$ (note that we assume $\eta_i' = \eta_o$). If decoupling is achieved, then $rank\ H_{zv'} = \eta_o$ since $H_{zv'}$ is nonsingular; hence, a necessary condition for decoupling is that $rank\ N_{12} = \eta_o$. But by (3.3.30), $P_{12} = N_{12} D_p^{-1}$ (i.e., $N_{12} = P_{12} D_p$) implies that $rank\ P_{12} = rank\ N_{12}$. Therefore, from now on we assume that

$$rank\ P_{12} = \eta_o \leq n_i \ . \tag{3.3.72}$$

In the case that P_{12} is square, (i.e., $\eta_o = n_i$ and hence, N_{12} is also square), condition (3.3.72) means that $\det P_{12} \neq 0$ (i.e., $\det N_{12} \neq 0$ because $\det P_{12} \neq 0$ and $\det D_p \in \mathbf{I}$).

In order to find compensators $\hat{C}$ that decouple the system $\Sigma(\hat{P}, \hat{C})$, we define two diagonal, nonsingular matrices Δ_L and Δ_R as follows:

Let $\Delta_{Lk} \in \mathbf{H}$ be a greatest-common-divisor (g.c.d.) of the elements of the k-th row of N_{12}. Let

$$\Delta_L := diag \begin{bmatrix} \Delta_{L1} & \cdots & \Delta_{L\eta_o} \end{bmatrix} ; \tag{3.3.73}$$

the diagonal matrix $\Delta_L \in \mathfrak{m}(\mathbf{H})$ is nonsingular since $\Delta_{Lk} \neq 0$ for $k = 1, \cdots, \eta_o$; furthermore, the diagonal entries Δ_{Lk} of Δ_L are unique within factors in $\mathbf{J}$. Let

$$N_{12} =: \Delta_L \hat{N}_{12} , \tag{3.3.74}$$

where the matrix $\hat{N}_{12} \in \mathrm{H}^{\eta_o \times n_i}$ has full normal row-rank since $rank\, N_{12} = \eta_o$ and $\det\Delta_L \neq 0$; therefore, $\hat{N}_{12}$ has a right-inverse denoted by $\hat{N}_{12}^I$; note that $\hat{N}_{12}^I \in \mathrm{m}(\mathrm{F})$. Write the ij-th entry of $\hat{N}_{12}^I$ as $\dfrac{m_{ij}}{d_{ij}}$, where m_{ij}, $d_{ij} \in \mathrm{H}$, $d_{ij} \neq 0$ and (m_{ij}, d_{ij}) is a coprime pair over H.

For $j = 1, \cdots, \eta_o$, let $\Delta_{Rj} \in \mathrm{H}$ be a least-common-multiple (l.c.m.) of the denominators of the entries in the j-th column of $\hat{N}_{12}^I$. Let

$$\Delta_R := diag \begin{bmatrix} \Delta_{R1} & \cdots & \Delta_{R\eta_o} \end{bmatrix} ; \tag{3.3.75}$$

the diagonal matrix $\Delta_R \in \mathrm{m}(\mathrm{H})$ is nonsingular since $\Delta_{Rj} \neq 0$ for $j = 1, \cdots, \eta_o$; furthermore, the diagonal entries Δ_{Rj} of Δ_R are unique within factors in J. (In the case that $\hat{N}_{12}^I \in \mathrm{m}(\mathrm{H})$, the denominators $d_{ij} \in \mathrm{J}$ and hence, without loss of generality, Δ_R is the identity matrix I_{η_o}.)

Now by definition, $\Delta_{Rj} = b_{ij}\, d_{ij}$ for some $b_{ij} \in \mathrm{H}$; therefore the ij-th entry of $\hat{N}_{12}^I \Delta_R$ is $\dfrac{m_{ij}}{d_{ij}} \Delta_{Rj} = m_{ij}\, b_{ij} \in \mathrm{H}$; hence,

$$\hat{N}_{12}^I \Delta_R \in \mathrm{m}(\mathrm{H}) . \tag{3.3.76}$$

Intuitively, if H is the ring $\mathrm{R_U}$ as in Section 2.2, then we can interpret the diagonal matrices Δ_L and Δ_R as follows: Since $rank\, N_{12} = \eta_o$ by assumption, $z \in \bar{\mathrm{U}}$ is a $\bar{\mathrm{U}}$–zero of N_{12} if and only if $rank\, N_{12}(z) < \eta_o$. Now Δ_{Lk} extracts the $\bar{\mathrm{U}}$–zeros that are common to all elements in the k-th row of N_{12}; Δ_L "book-keeps" the $\bar{\mathrm{U}}$–zeros of $P_{12} = N_{12} D_p^{-1}$ that appear in each entry of some row of N_{12}. Clearly, P_{12} may have other $\bar{\mathrm{U}}$–zeros that cannot be extracted by Δ_L; these $\bar{\mathrm{U}}$–zeros are the $\bar{\mathrm{U}}$–zeros of $\hat{N}_{12}$ (equivalently, the $\bar{\mathrm{U}}$–poles of $\hat{N}_{12}^I$). Suppose for simplicity that P_{12} is square: Multiplying $\hat{N}_{12}$ by the diagonal matrix Δ_R makes $\hat{N}_{12}^{-1} \Delta_R$ H–stable, i.e., cancels these $\bar{\mathrm{U}}$–poles. Let $s \in \bar{\mathrm{U}}$ be a zero of Δ_R (hence a $\bar{\mathrm{U}}$–zero of $\det\hat{N}_{12}$); the multiplicity of $s \in \bar{\mathrm{U}}$ in $\det\Delta_R$ may exceed its multiplicity in

$\det\hat{N}_{12}$. If $\det\hat{N}_{12} \in \mathrm{H}^{\eta_o \times \eta_o}$ has n zeros at $s \in \bar{\mathcal{U}}$, then $\det\Delta_R$ has *at most* n^{η_o} zeros at $s \in \bar{\mathcal{U}}$; so Δ_R has at most as many $\bar{\mathcal{U}}$–zeros as $(\det\hat{N}_{12}) I_{\eta_o}$.

Definition 3.3.14. (Achievable diagonal input-output map $H_{zv'}$)

The set

$$\hat{\mathrm{A}}_{zv'}(\hat{P}) := \{\ H_{zv'}\ \mid\ \hat{C}\ \text{H–}\textit{stabilizes}\ \hat{P}\ \text{and the map}\ H_{zv'}\ \text{is}\ \textit{diagonal}\ \text{and}\ \textit{nonsingular}\ \} \tag{3.3.77}$$

is called the set of *all achievable diagonal nonsingular input-output maps* $H_{zv'} : v' \mapsto z$.

□

Clearly, $\hat{\mathrm{A}}_{zv'}(\hat{P})$ is a subset of the achievable $v' \mapsto z$ maps in $\hat{\mathrm{A}}(\hat{P})$ because $\hat{C}$ must be a H–stabilizing compensator; in other words, $\hat{\mathrm{A}}_{zv'}(\hat{P})$ is the set of all $N_{12} Q_{21} \in \mathrm{m(H)}$ that are *diagonal* and *nonsingular.* Thus we must choose the parameter $Q_{21} \in \mathrm{m(H)}$ so that $N_{12} Q_{21}$ is diagonal and nonsingular (see equation (3.3.71)). The "minimal" restriction on Q_{21} to achieve diagonal $H_{zv'}$ is given in Theorem 3.3.15 below:

Theorem 3.3.15. (Class of all achievable diagonal $H_{zv'}$)

Let $\hat{P} \in \mathrm{G}^{(\eta_o+n_o)\times(\eta_i+n_i)}$ be Σ–admissible. Suppose that $\eta_i' = \eta_o$. Without loss of generality, assume that an r.c.f.r. of $\hat{P}$ is given by (3.3.30) and an l.c.f.r. of $\hat{P}$ is given by (3.3.31). Let $P_{12} \in \mathrm{G}^{\eta_o \times n_i}$ have full normal row-rank; so *rank* $N_{12} = \eta_o$. Under these assumptions, the set $\hat{\mathrm{A}}_{zv'}(\hat{P})$ of all achievable diagonal nonsingular input-output maps $H_{zv'}$ is given by

$$\hat{\mathrm{A}}_{zv'}(\hat{P}) = \{\ \Delta_L\, \Delta_R\, \hat{Q}_{21}\ \mid\ \hat{Q}_{21} \in \mathrm{H}^{\eta_o \times \eta_o}\ \text{is diagonal and nonsingular}\ \}, \tag{3.3.78}$$

where Δ_L and Δ_R are the diagonal, nonsingular matrices defined by equations (3.3.73) and (3.3.75), respectively.

Comment 3.3.16

(i) The square and diagonal input output map $H_{zv'} = \Delta_L \, \Delta_R \, \hat{Q}_{21} \in \mathbf{H}^{n_o \times n_o}$ is achieved by choosing the compensator parameter Q_{21} as

$$Q_{21} = \hat{N}_{12}^{I} \, \Delta_R \, \hat{Q}_{21} \ , \tag{3.3.79}$$

where $\hat{Q}_{21} \in \mathbf{H}^{n_o \times n_o}$ is diagonal and nonsingular. By equation (3.3.76), if Q_{21} is chosen as in (3.3.79), then $Q_{21} \in \mathrm{m}(\mathbf{H})$. Therefore, to achieve diagonalization, from the set $\hat{S}(\hat{P})$ of all $\mathbf{H}$–stabilizing compensators $\hat{C}$, we must choose $C_{21} = \tilde{D}_c^{-1} Q_{21} = (V_p - Q \tilde{N}_p)^{-1} Q_{21}$ as

$$C_{21} = (V_p - Q \tilde{N}_p)^{-1} \hat{N}_{12}^{I} \, \Delta_R \, \hat{Q}_{21} \ , \tag{3.3.80}$$

where $\hat{Q}_{21} \in \mathbf{H}^{n_o \times n_o}$ is a diagonal, nonsingular matrix and $Q \in \mathbf{H}^{n_i \times n_o}$ is chosen such that $\det(V_p - Q \tilde{N}_p) \in \mathbf{I}$. Note that in equation (3.3.80), the (matrix-) parameter $Q \in \mathbf{H}^{n_i \times n_o}$ is *not* used in diagonalizing the I/O map $H_{zv'}$; in the case that $P \in \mathrm{m}(\mathbf{G}_s)$, $Q \in \mathrm{m}(\mathbf{H})$ is completely free since $\det(V_p - Q \tilde{N}_p) \in \mathbf{I}$ for all $Q \in \mathrm{m}(\mathbf{H})$. The other compensator parameters Q_{11} and Q_{12} are not used in diagonalizing the map $H_{zv'}$ either.

Note that the parameter Q_{21} is restricted to be $\hat{N}_{12}^{I} \Delta_R \hat{Q}_{21}$ for decoupling and hence, can no longer be assigned arbitrarily in order to meet other design specifications; the only freedom left is the diagonal nonsingular matrix $\hat{Q}_{21} \in \mathrm{m}(\mathbf{H})$, in addition to the parameters Q_{11}, Q_{12} and Q .

(ii) If $\mathbf{H}$ is the ring $\mathbf{R_U}$ as in Section 2.2 and if P_{12} is square for simplicity, then the "cost" of diagonalizing the map $H_{zv'}$ is that the number of its $\bar{\mathcal{U}}$–zeros are increased. Since Δ_L is a factor of N_{12} , $H_{zv'}$ must have zeros at the $\bar{\mathcal{U}}$–zeros of Δ_L ; the multiplicity of a $\bar{\mathcal{U}}$–zero of $H_{zv'}$ may be larger than its multiplicity in $\det N_{12}$ due to the factor Δ_R . If Δ_L is the only source of $\bar{\mathcal{U}}$–zeros of P_{12} (equivalently, if $\hat{N}_{12}^{-1} \in \mathrm{m}(\mathbf{H})$) and if $\hat{Q}_{21}$ is chosen so that it has no $\bar{u}$–zeros, then the $\bar{u}$–zeros of the diagonal $H_{zv'}$ have the same multiplicity as in $\det N_{12}$ since in that case $\Delta_R = I_{\eta_o}$.

(iii) Although we chose to diagonalize the map $H_{zv'}$, we could also diagonalize $H_{yv'} : v' \mapsto y$, the map from the same external-input v' to the measured output y of $\hat{P}$ (y is the output used in feedback). In that case, assuming that $\eta_i' = n_o$ and that $rank\ P = n_o \le n_i$, we would define Δ_{Rp}, Δ_{Lp} and $\hat{N}_p$ from N_p as we did above to obtain Δ_L, Δ_R and $\hat{N}_{12}$ from N_{12}; the set of all achievable diagonal nonsingular maps $H_{yv'}$ would then be $\hat{A}_{yv'}(\hat{P})$, where

$$\hat{A}_{yv'}(\hat{P}) = \{\ \Delta_{Lp}\ \Delta_{Rp}\ \hat{Q}_{21}\ \mid\ \hat{Q}_{21} \in \mathrm{m}(\mathrm{H}) \text{ is diagonal and nonsingular }\}\ . \quad (3.3.81)$$

The compensator parameter Q_{21} must then be chosen as

$$\hat{N}_p^I\ \Delta_{Rp}\ \hat{Q}_{21}\ , \qquad (3.3.82)$$

where $\hat{N}_p^I$ is the right-inverse of $\hat{N}_p$. □

Proof of Theorem 3.3.15

The map $H_{zv'}$ is an achievable map of $\Sigma(\hat{P}, \hat{C})$ if and only if $H_{zv'} = N_{12}\,Q_{21}$ for some $Q_{21} \in \mathrm{m}(\mathrm{H})$. By equation (3.3.74),

$$H_{zv'} = N_{12}\,Q_{21} = \Delta_L\,\hat{N}_{12}\,Q_{21} \qquad (3.3.83)$$

for some $Q_{21} \in \mathrm{m}(\mathrm{H})$. Now $H_{zv'} \in \mathrm{m}(\mathrm{H})$ is diagonal and nonsingular if and only if $Q_{21} \in \mathrm{m}(\mathrm{H})$ is such that $\Delta_L\,\hat{N}_{12}\,Q_{21}$ is diagonal and nonsingular. Choose Q_{21} as in equation (3.3.79); then by equation (3.3.76), $Q_{21} \in \mathrm{m}(\mathrm{H})$. Clearly, $H_{zv'} = \Delta_L\,\Delta_R\,\hat{Q}_{21}$ is an achievable diagonal nonsingular map.

Now if $H_{zv'}$ is a given diagonal map achieved by $\Sigma(\hat{P}, \hat{C})$, then by equations (3.3.71)nd (3.3.74), Δ_L is clearly a factor of $H_{zv'}$. Now suppose, for a contradiction, that $Q_{21} = \hat{N}_{12}^I\,\hat{\Delta}_R\,\hat{Q}_{21}$, i.e., that

$$H_{zv'} = \Delta_L \hat{\Delta}_R \hat{Q}_{21} , \tag{3.3.84}$$

where all (diagonal) entries of $\hat{\Delta}_R$ are the same as those of Δ_R except the j-th entry, which is a proper factor of Δ_{Rj} , i.e., for some $\delta_j \notin \mathbf{J}$,

$$\Delta_{Rj} = \hat{\Delta}_{Rj} \, \delta_j . \tag{3.3.85}$$

Since for $i = 1 , \cdots , n_i$, Δ_{Rj} is a l.c.m. of d_{ij} , some denominator, say the k-th row j-th column denominator d_{kj} has that factor δ_j , i.e., $d_{kj} = \delta_j \hat{d}_{kj}$. The kj-th entry of Q_{21} is then $\frac{m_{kj}}{d_{kj}} \hat{\Delta}_{Rj} q_j$, where q_j is the j-th (diagonal) entry of $\hat{Q}_{21}$. Since δ_j is not a factor of $\hat{\Delta}_{Rj}$ and since (m_{kj} , d_{kj}) is a coprime pair, the only way that the kj-th entry of Q_{21} will be in $\mathbf{H}$ is if $q_j = \delta_j q_j'$ for some $q_j' \in \mathbf{H}$; $\hat{Q}_{21}$ then becomes $diag \left[1 \cdots \delta_j \cdots 1 \right] \hat{Q}'_{21}$. Therefore, $H_{zv'} = \Delta_L \hat{\Delta}_R \, diag \left[1 \cdots \delta_j \cdots 1 \right] \hat{Q}'_{21} = \Delta_L \Delta_R \hat{Q}'_{21}$ for some $\hat{Q}'_{21} \in \mathfrak{m}(\mathbf{H})$. □

We conclude this chapter by applying some of the results of Sections 3.2 and 3.3 to a Σ–admissible plant $\hat{P}$ whose entries are proper rational functions.

Example 3.3.17

Let $\mathbf{H}$ be the ring $\mathbf{R_u}$ as in Section 2.2. Consider the two (vector-) input two (vector-) output plant

$$\hat{P} = \begin{bmatrix} P_{11} & P_{12} \\ P_{21} & P \end{bmatrix} = \begin{bmatrix} \tilde{C} \\ \bar{C} \end{bmatrix} (s I_n - \bar{A})^{-1} \begin{bmatrix} \tilde{B} & \bar{B} \end{bmatrix} + \begin{bmatrix} 0 & 0 \\ 0 & \bar{E} \end{bmatrix} \in \mathfrak{m}(\mathbb{R}_p(s)),$$

where $\bar{A} \in \mathbb{R}^{n \times n}$, $\bar{B} \in \mathbb{R}^{n \times n_i}$, $\bar{C} \in \mathbb{R}^{n_o \times n}$, $\bar{E} \in \mathbb{R}^{n_o \times n_i}$, $\tilde{B} \in \mathbb{R}^{n \times \eta_i}$ and $\tilde{C} \in \mathbb{R}^{\eta_o \times n}$. Let $(\bar{A} , \bar{B})$ be $\bar{\mathcal{U}}$–stabilizable, $(\bar{C} , \bar{A})$ be $\bar{\mathcal{U}}$–detectable; let $a \in \mathbb{R}$ and the matrices $K \in \mathbb{R}^{n_i \times n}$, $F \in \mathbb{R}^{n \times n_o}$, $A_k \in \mathbf{R_u}^{n \times n}$, $A_f \in \mathbf{R_u}^{n \times n}$ be defined as in Example 2.4.3. Clearly,

$$(N_{pr}, D, N_{pl}, G) := ((s+a)^{-1}\tilde{C}, (s+a)^{-1}(sI_n - \bar{A}), \bar{B}, \bar{E})$$

is a b.c.f.r. of $P \in \mathbb{R}_p(s)^{n_o \times n_i}$.

It is easy to check that $\hat{P}$ is Σ–admissible by testing conditions (3.3.32)-(3.3.34) of Theorem 3.3.9: From the r.c.f.r. and l.c.f.r. of P given by (2.4.18) and (2.4.19), $D_p = (I_{n_i} - K A_k \bar{B})$ and $\tilde{D}_p = (I_{n_o} - \tilde{C} A_f F)$; from the generalized Bezout identity (2.4.17), $U_p = K A_f F$. Now condition (3.3.32) is satisfied since $A_k, A_f \subset \mathrm{m}(\mathrm{R_u})$ implies that

$$\hat{N}_{11} = P_{11} - P_{12} D_p U_p P_{21} = \tilde{C} A_k (I_n + \bar{B} K A_f) \tilde{B}$$

$$= \tilde{C} A_k (sI_n - \bar{A} + F\bar{C} + \bar{B}K) A_f \tilde{B} \in \mathrm{m}(\mathrm{R_u});$$

condition (3.3.33) is satisfied since

$$N_{12} = P_{12} D_p = \tilde{C} A_k \bar{B} \in \mathrm{m}(\mathrm{R_u});$$

condition (3.3.34) is satisfied since

$$\tilde{N}_{21} = \tilde{D}_p P_{21} = \tilde{C} A_f \tilde{B} \in \mathrm{m}(\mathrm{R_u}).$$

We can now write a right-coprime factorization of $\hat{P}$ using (3.3.30) as: $\hat{P} = N_{\hat{P}} D_{\hat{P}}^{-1}$

$$= \begin{bmatrix} \tilde{C} A_k (I_n + \bar{B} K A_f) \tilde{B} & \tilde{C} A_k \bar{B} \\ (I_{n_o} + \bar{C} A_k F - \bar{E} K A_k F) \bar{C} A_f \tilde{B} & \bar{C} A_k \bar{B} + \bar{E}(I_{n_i} - K A_k \bar{B}) \end{bmatrix} \begin{bmatrix} I_{\eta_i} & 0 \\ -K A_k F \bar{C} A_f \tilde{B} & I_{n_i} - K A_k \bar{B} \end{bmatrix}^{-1}.$$

The class of all $\mathrm{R_u}$–stabilizing compensators for this $\hat{P}$ can be found from (3.3.53)-(3.3.54) using (2.4.17), (2.4.18), (2.4.19). The set of all achievable maps $H_{zv'}$ from the external input v' to the actual-output z of $\hat{P}$ is given by:

$$\{ H_{zv'} = \tilde{C} A_k \bar{B} Q_{21} \mid Q_{21} \in \mathrm{R_u}^{n_i \times \eta_i'} \}.$$

Another I/O map of interest is the map $H_{zv} : v \mapsto z$, which is often encountered in H_∞ optimal design problems; the set of all achievable maps H_{zv} for the plant in this example is:

$$\{ H_{zv} = \tilde{C} A_k (I_n + \bar{B} K A_f \tilde{B}) - \tilde{C} A_k \bar{B} Q \tilde{C} A_f \tilde{B} \mid$$

$$Q \in \mathrm{R_u}^{n_i \times n_o}, \ \det(V_p - Q \tilde{N}_p) \in \mathrm{I} \}. \square$$

Chapter 4

DECENTRALIZED CONTROL SYSTEMS

4.1 INTRODUCTION

In Chapter Three we studied two system configurations: $\mathbf{S}(P, C)$ and $\Sigma(\hat{P}, \hat{C})$; these systems put no constraints on the structure of the feedback compensator. We now study the consequences of restricting the compensator to be block-diagonal.

Restrictions on the feedback compensator structure are often encountered in large scale systems. These systems have several local control stations; each local compensator observes only the corresponding (local) outputs. Such *decentralized* control of systems results in a block-diagonal compensator-matrix structure.

In this chapter, using the completely general algebraic framework of Chapter Two, we obtain necessary and sufficient conditions on a plant P for stabilizability by a decentralized dynamic compensator. Decentralized stabilizability conditions turn out to be that certain canonical forms resembling the Smith-form must be satisfied by the coprime factorizations of the plant P. When the compensator structure is required to be block-diagonal as in decentralized output-feedback, finding the class of all stabilizing decentralized compensators is complicated; the task is to find *structured* generalized Bezout identities associated with coprime factorizations plant P. For plants that satisfy decentralized stabilizability conditions, we parametrize the class of all decentralized stabilizing compensators.

The two-channel decentralized control system $\mathbf{S}(P, C_d)$ is studied in detail in sections 4.2 through 4.4; the results are extended to the m-channel decentralized feedback control system $\mathbf{S}(P, C_d)_m$ in Section 4.5. In Section 4.2, following the analysis using coprime factorizations of P and C_d, Theorem 4.2.5 gives four equivalent necessary and sufficient conditions for the H–stability of $\mathbf{S}(P, C_d)$. Section 4.3 contains two important results: Theorem 4.3.3 (Conditions on $P = N_p D_p^{-1} = \tilde{D}_p^{-1} \tilde{N}_p$ for decentralized H–stabilizability) and Theorem 4.3.5 (Class of all decentralized H–stabilizing compensators in $\mathbf{S}(P, C_d)$). In Section 4.4 it is shown that the decentralized H–stabilizability conditions of Theorem 4.3.3 are equivalent to the condition that the proper rational matrix P has no decentralized fixed-eigenvalues in the undesirable region $\bar{\mathcal{U}}$; these conditions are shown to be equivalent to certain rank tests on coprime factorizations of P in Theorem 4.4.4 (Rank tests on $P = N_p D_p^{-1} = \tilde{D}_p^{-1} \tilde{N}_p$ for decentralized fixed-eigenvalues and H–stabilizability). Similar rank tests on state-space realizations of P are given in Remark 4.4.6. It is interesting to note the relationship between decentralized fixed-eigenvalues and transmission-zeros of the plant (Comment 4.4.8). Section 4.4.9 gives an algorithm for designing a two-channel decentralized H–stabilizing compensator in the case that P has strictly-proper rational function entries; a simple example is also included (Example 4.4.10). In Section 4.5, Theorems 4.5.4 and 4.5.5 are extensions of the main results of Section 4.3 to the m-channel case; Comment 4.5.6 summarizes the rational functions case, which was explained in detail for two-channels in Section 4.4. Finally in Section 4.5.7 we see that the input-output maps achieved by the H–stabilized decentralized control system $\mathbf{S}(P, C_d)_m$ (and the system $\mathbf{S}(P, C_d)$ in the case that $m = 2$) are *not affine* in the (matrix-) parameters that describe all decentralized H–stabilizing compensators.

4.2 TWO-CHANNEL DECENTRALIZED CONTROL SYSTEM

In this section we consider the linear, time-invariant, two-channel decentralized feedback system $\mathbf{S}(P, C_d)$ shown in Figure 4.1, where $P : \begin{bmatrix} e_1 \\ e_2 \end{bmatrix} \mapsto \begin{bmatrix} y_1 \\ y_2 \end{bmatrix}$ represents the plant and $C : \begin{bmatrix} e_1' \\ e_2' \end{bmatrix} \mapsto \begin{bmatrix} y_1' \\ y_2' \end{bmatrix}$ represents the compensator. The externally applied inputs are denoted by $\bar{u} := \begin{bmatrix} u_1 \\ u_2 \\ u_1' \\ u_2' \end{bmatrix}$, the plant and the compensator outputs are denoted by $\bar{y} := \begin{bmatrix} y_1 \\ y_2 \\ y_1' \\ y_2' \end{bmatrix}$; the closed-loop input-output map of $\mathbf{S}(P, C_d)$ is denoted by $H_{\overline{yu}} : \bar{u} \mapsto \bar{y}$.

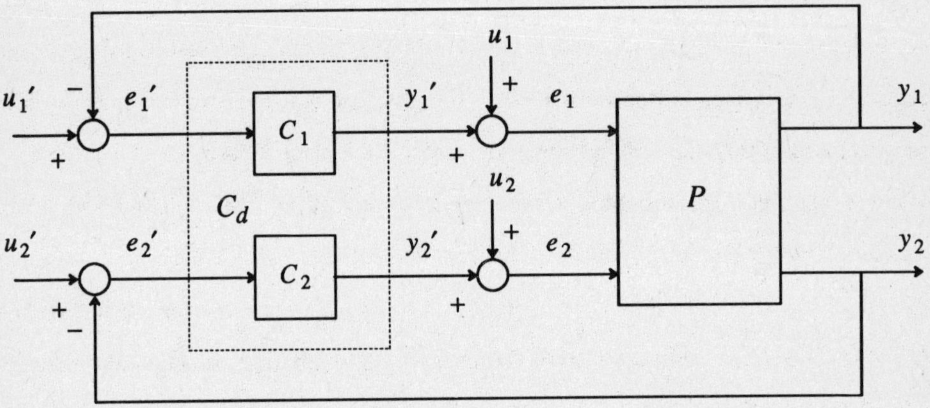

Figure 4.1. The two-channel decentralized control system $\mathbf{S}(P, C_d)$.

4.2.1. Assumptions on $\mathbf{S}(P, C_d)$

(i) The two-channel plant $P \in \mathbf{G}^{n_o \times n_i}$, where

$$P = \begin{bmatrix} P_{11} & P_{12} \\ P_{21} & P_{22} \end{bmatrix},$$

$$P_{11} \in \mathbf{G}^{n_{o1} \times n_{i1}}, \; P_{12} \in \mathbf{G}^{n_{o1} \times n_{i2}}, \; P_{21} \in \mathbf{G}^{n_{o2} \times n_{i1}}, \; P_{22} \in \mathbf{G}^{n_{o2} \times n_{i2}},$$

$$n_o =: n_{o1} + n_{o2}, \quad n_i =: n_{i1} + n_{i2}.$$

(ii) The decentralized compensator $C_d \in \mathbf{G}^{n_i \times n_o}$, where

$$C_d = \begin{bmatrix} C_1 & 0 \\ 0 & C_2 \end{bmatrix}, \; C_1 \in \mathbf{G}^{n_{i1} \times n_{o1}}, \; C_2 \in \mathbf{G}^{n_{i2} \times n_{o2}}.$$

(iii) The system $\mathbf{S}(P, C_d)$ is well-posed; equivalently, the closed-loop input-output map $H_{\overline{yu}} : \bar{u} \mapsto \bar{y}$ is in $\mathfrak{m}(\mathbf{G})$. □

Note that whenever P satisfies Assumption 4.2.1 **(i)**, it has an r.c.f.r., denoted by (N_p, D_p), an l.c.f.r., denoted by $(\tilde{D}_p, \tilde{N}_p)$ and a b.c.f.r., denoted by (N_{pr}, D, N_{pl}, G), where the numerator and the denominator matrices can be partitioned as follows: In the r.c.f.r. (N_p, D_p),

$$N_p =: \begin{bmatrix} N_{p1} \\ N_{p2} \end{bmatrix} \in \mathbf{H}^{n_o \times n_i}, \quad D_p =: \begin{bmatrix} D_{p1} \\ D_{p2} \end{bmatrix} \in \mathbf{H}^{n_i \times n_i}, \tag{4.2.1}$$

where $N_{p1} \in \mathbf{H}^{n_{o1} \times n_i}$, $N_{p2} \in \mathbf{H}^{n_{o2} \times n_i}$, $D_{p1} \in \mathbf{H}^{n_{i1} \times n_i}$, $D_{p2} \in \mathbf{H}^{n_{i2} \times n_i}$.

In the l.c.f.r. $(\tilde{D}_p, \tilde{N}_p)$,

$$\tilde{D}_p =: \begin{bmatrix} \tilde{D}_{p1} & \tilde{D}_{p2} \end{bmatrix} \in \mathbf{H}^{n_o \times n_o}, \quad \tilde{N}_p =: \begin{bmatrix} \tilde{N}_{p1} & \tilde{N}_{p2} \end{bmatrix} \in \mathbf{H}^{n_o \times n_i}, \tag{4.2.2}$$

where $\tilde{D}_{p1} \in \mathbf{H}^{n_o \times n_{o1}}$, $\tilde{D}_{p2} \in \mathbf{H}^{n_o \times n_{o2}}$, $\tilde{N}_{p1} \in \mathbf{H}^{n_o \times n_{i1}}$, $\tilde{N}_{p2} \in \mathbf{H}^{n_o \times n_{i2}}$.

In the b.c.f.r. (N_{pr}, D, N_{pl}, G),

$$N_{pr} =: \begin{bmatrix} N_{pr1} \\ N_{pr2} \end{bmatrix} \in \mathbf{H}^{n_o \times n}, \quad N_{pl} =: \begin{bmatrix} N_{pl1} & N_{pl2} \end{bmatrix} \in \mathbf{H}^{n \times n_i}, \quad G =: \begin{bmatrix} G_{11} & G_{12} \\ G_{21} & G_{22} \end{bmatrix}, \tag{4.2.3}$$

where $N_{pr1} \in \mathbf{H}^{n_{o1} \times n}$, $N_{pr2} \in \mathbf{H}^{n_{o2} \times n}$, $N_{pl1} \in \mathbf{H}^{n \times n_{i1}}$, $N_{pl2} \in \mathbf{H}^{n \times n_{i2}}$, $G_{11} \in \mathbf{H}^{n_{o1} \times n_{i1}}$, $G_{12} \in \mathbf{H}^{n_{o1} \times n_{i2}}$, $G_{21} \in \mathbf{H}^{n_{o2} \times n_{i1}}$, $G_{22} \in \mathbf{H}^{n_{o2} \times n_{i2}}$; $D \in \mathbf{H}^{n \times n}$.

The generalized Bezout identity (2.3.12) for the doubly-coprime pair

$((N_p, D_p), (\tilde{D}_p, \tilde{N}_p))$ is satisfied for some $V_p, U_p, \tilde{V}_p, \tilde{U}_p \in \mathbf{m}(\mathrm{H})$. For the b.c.f.r. (N_{pr}, D, N_{pl}, G) of P we have the two generalized Bezout identities (2.3.13) and (2.3.14), partitioned as follows: For the r.c. pair (N_{pr}, D), there are matrices $V_{pr}, U_{pr}, \tilde{X}, \tilde{Y}, \tilde{U}, \tilde{V} \in \mathbf{m}(\mathrm{H})$ such that

$$\begin{bmatrix} V_{pr} & U_{pr} \\ -\tilde{X} & \tilde{Y} \end{bmatrix} \begin{bmatrix} D & -\tilde{U} \\ N_{pr} & \tilde{V} \end{bmatrix} =: \begin{bmatrix} V_{pr} & U_{pr1} & U_{pr2} \\ -\tilde{X} & \tilde{Y}_1 & \tilde{Y}_2 \end{bmatrix} \begin{bmatrix} D & -\tilde{U} \\ N_{pr1} & \tilde{V}_1 \\ N_{pr2} & \tilde{V}_2 \end{bmatrix}$$

$$= \begin{bmatrix} I_n & 0 \\ 0 & I_{n_o} \end{bmatrix} ; \tag{4.2.4}$$

for the l.c. pair (D, N_{pl}), there are matrices $V_{pl}, U_{pl}, X, Y, U, V \in \mathbf{m}(\mathrm{H})$ such that

$$\begin{bmatrix} D & -N_{pl} \\ U & V \end{bmatrix} \begin{bmatrix} V_{pl} & X \\ -U_{pl} & Y \end{bmatrix} =: \begin{bmatrix} D & -N_{pl1} & -N_{pl2} \\ U & V_1 & V_2 \end{bmatrix} \begin{bmatrix} V_{pl} & X \\ -U_{pl1} & Y_1 \\ -U_{pl2} & Y_2 \end{bmatrix}$$

$$= \begin{bmatrix} I_n & 0 \\ 0 & I_{n_i} \end{bmatrix} . \tag{4.2.5}$$

If C_d satisfies Assumption 4.2.1 (ii), then C_d has an l.c.f.r., denoted by $(\tilde{D}_c, \tilde{N}_c)$ and an r.c.f.r., denoted by (N_c, D_c), where $\tilde{D}_c \in \mathrm{H}^{n_i \times n_i}$, $\tilde{N}_c \in \mathrm{H}^{n_i \times n_o}$, $N_c \in \mathrm{H}^{n_i \times n_o}$, $D_c \in \mathrm{H}^{n_o \times n_o}$. Note that $(\tilde{D}_c, \tilde{N}_c)$ is an l.c.f.r. of C_d if and only if $(\tilde{D}_{c1}, \tilde{N}_{c1})$ is an l.c.f.r. of C_1 and $(\tilde{D}_{c2}, \tilde{N}_{c2})$ is an l.c.f.r. of C_2, where $\tilde{D}_{c1} \in \mathrm{H}^{n_{i1} \times n_{i1}}$, $\tilde{D}_{c2} \in \mathrm{H}^{n_{i2} \times n_{i2}}$, $\tilde{N}_{c1} \in \mathrm{H}^{n_{i1} \times n_{o1}}$, $\tilde{N}_{c2} \in \mathrm{H}^{n_{i2} \times n_{o2}}$ are as in (4.2.6) below:.

$$\tilde{D}_c = \begin{bmatrix} \tilde{D}_{c1} & 0 \\ 0 & \tilde{D}_{c2} \end{bmatrix}, \quad \tilde{N}_c = \begin{bmatrix} \tilde{N}_{c1} & 0 \\ 0 & \tilde{N}_{c2} \end{bmatrix} ; \tag{4.2.6}$$

Similarly, (N_c, D_c) is an r.c.f.r. of C_d if and only if (N_{c1}, D_{c1}) is an r.c.f.r. of C_1 and (N_{c2}, D_{c2}) is an r.c.f.r. of C_2, where $D_{c1} \in \mathrm{H}^{n_{o1} \times n_{o1}}$, $D_{c2} \in \mathrm{H}^{n_{o2} \times n_{o2}}$, $N_{c1} \in \mathrm{H}^{n_{i1} \times n_{o1}}$, $N_{c2} \in \mathrm{H}^{n_{i2} \times n_{o2}}$ are as in (4.2.7) below:

$$D_c = \begin{bmatrix} D_{c1} & 0 \\ 0 & D_{c2} \end{bmatrix}, \quad N_c = \begin{bmatrix} N_{c1} & 0 \\ 0 & N_{c2} \end{bmatrix}; \tag{4.2.7}$$

4.2.2. Closed-loop input-output maps of $\mathrm{S}(P, C_d)$

Let Assumptions 4.2.1 hold; the closed-loop I/O map $H_{\overline{yu}}$ of the system $\mathrm{S}(P, C_d)$ is given in terms of $(I_{n_i} + C_d P)^{-1}$ in equation (4.2.8) and in terms of $(I_{n_o} + P C_d)^{-1}$ in equation (4.2.9) below:

$$H_{\overline{yu}} = \begin{bmatrix} P(I_{n_i} + C_d P)^{-1} & P(I_{n_i} + C_d P)^{-1} C_d \\ (I_{n_i} + C_d P)^{-1} - I_{n_i} & (I_{n_i} + C_d P)^{-1} C_d \end{bmatrix}; \tag{4.2.8}$$

$$H_{\overline{yu}} = \begin{bmatrix} (I_{n_o} + P C_d)^{-1} P & (I_{n_o} + P C_d)^{-1} P C_d \\ -C_d (I_{n_o} + P C_d)^{-1} P & C_d (I_{n_o} + P C_d)^{-1} \end{bmatrix}. \tag{4.2.9}$$

Note that equations (4.2.8)-(4.3.9) are the same as equations (3.2.6)-(3.2.7), where C_d is replaced by C.

Definition 4.2.3. **(H–stability of $\mathrm{S}(P, C_d)$)**

The system $\mathrm{S}(P, C_d)$ is said to be H–*stable* iff $H_{\overline{yu}} \in \mathrm{m}(\mathrm{H})$. □

4.2.4. Analysis (Descriptions of $S(P, C_d)$ using coprime factorizations)

We now analyze the decentralized feedback system $S(P, C_d)$ using coprime factorizations over $\mathbf{m}(\mathbf{H})$ of the plant and the compensator transfer matrices.

Assumptions 4.2.1 hold throughout this analysis.

(i) Analysis of $S(P, C_d)$ with $P = N_p D_p^{-1}$ and $C_d = \tilde{D}_c^{-1} \tilde{N}_c$

Let (N_p, D_p) be any r.c.f.r. of $P \in \mathbf{m}(\mathbf{G})$, partitioned as in equation (4.2.1) and let $(\tilde{D}_c, \tilde{N}_c)$ be any l.c.f.r. of $C_d \in \mathbf{m}(\mathbf{G})$, partitioned as in equation (4.2.6). The system $S(P, C_d)$ in Figure 4.1 can be redrawn as in Figure 4.2 below, where $P = N_p D_p^{-1}$, $C_1 = \tilde{D}_{c1}^{-1} \tilde{N}_{c1}$ and $C_2 = \tilde{D}_{c2}^{-1} \tilde{N}_{c2}$; note that $D_p\, \xi_p = \begin{bmatrix} e_1 \\ e_2 \end{bmatrix}$, $\begin{bmatrix} y_1 \\ y_2 \end{bmatrix} = \begin{bmatrix} N_{p1} \\ N_{p2} \end{bmatrix} \xi_p$, where ξ_p denotes the *pseudo-state* of P.

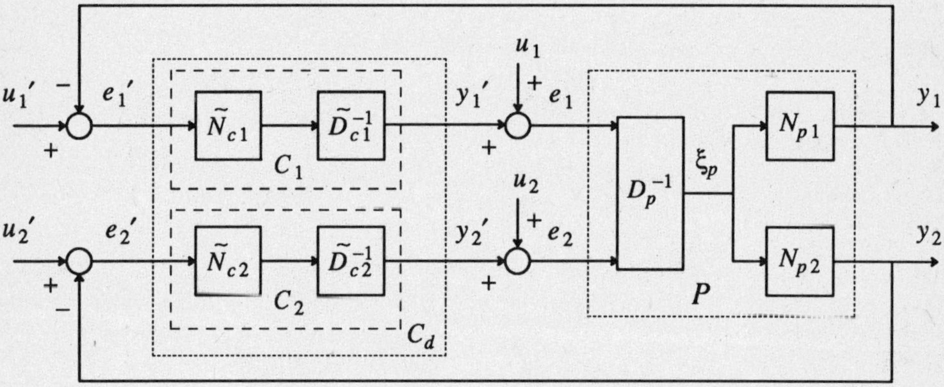

Figure 4.2. $S(P, C_d)$ with $P = N_p D_p^{-1}$ and $C_d = \tilde{D}_c^{-1} \tilde{N}_c$.

The system $S(P, C_d)$ is then described by equations (4.2.10)-(4.2.11) below:

$$\begin{bmatrix} \tilde{D}_{c1} D_{p1} + \tilde{N}_{c1} N_{p1} \\ \tilde{D}_{c2} D_{p2} + \tilde{N}_{c2} N_{p2} \end{bmatrix} \xi_p = \begin{bmatrix} \tilde{D}_{c1} & 0 & \tilde{N}_{c1} & 0 \\ 0 & \tilde{D}_{c2} & 0 & \tilde{N}_{c2} \end{bmatrix} \begin{bmatrix} u_1 \\ u_2 \\ u_1' \\ u_2' \end{bmatrix}, \quad (4.2.10)$$

$$\begin{bmatrix} N_{p1} \\ N_{p2} \\ D_{p1} \\ D_{p2} \end{bmatrix} \xi_p = \begin{bmatrix} y_1 \\ y_2 \\ y_1' \\ y_2' \end{bmatrix} - \begin{bmatrix} 0 & 0 & 0 & 0 \\ 0 & 0 & 0 & 0 \\ -I_{n_{i1}} & 0 & 0 & 0 \\ 0 & -I_{n_{i2}} & 0 & 0 \end{bmatrix} \begin{bmatrix} u_1 \\ u_2 \\ u_1' \\ u_2' \end{bmatrix}. \quad (4.2.11)$$

Equations (4.2.10)-(4.2.11) are of the form

$$D_{H1} \xi_p = N_{HL1} \bar{u}$$

$$N_{HR1} \xi_p = \bar{y} - G_{H1} \bar{u} .$$

By Lemma 2.3.2, performing elementary row operations over $\mathbf{m}(\mathbf{H})$ on the matrix $\begin{bmatrix} D_{H1} \\ N_{HR1} \end{bmatrix}$ and elementary column operations over $\mathbf{m}(\mathbf{H})$ on the matrix $\begin{bmatrix} N_{HL1} & D_{H1} \end{bmatrix}$, we conclude that $(N_{HR1}, D_{H1}, N_{HL1})$ is a b.c. triple. Since $D_{H1}, G_{H1} \in \mathbf{m}(\mathbf{H})$, it follows from Comment 2.4.7 (i) that

$$H_{\bar{y}\bar{u}} = N_{HR1} D_{H1}^{-1} N_{HL1} + G_{H1} \in \mathbf{m}(\mathbf{G}) \quad (4.2.12)$$

if and only if $\det D_{H1} \in \mathbf{I}$. Since Assumption 4.2.1 (iii) holds, the system $\mathbf{S}(P, C_d)$ is well-posed; therefore $\det D_{H1} \in \mathbf{I}$. Consequently, $(N_{HR1}, D_{H1}, N_{HL1}, G_{H1})$ is a b.c.f.r. of $H_{\bar{y}\bar{u}}$ and hence, $\det D_{H1}$ is a characteristic determinant of $H_{\bar{y}\bar{u}}$.

(ii) Analysis of $\mathbf{S}(P, C_d)$ with $P = \tilde{D}_p^{-1} \tilde{N}_p$ and $C_d = N_c D_c^{-1}$

Let $(\tilde{D}_p, \tilde{N}_p)$ be any l.c.f.r. of $P \in \mathbf{m}(\mathbf{G})$, partitioned as in equation (4.2.2) and let (N_c, D_c) be any r.c.f.r. of $C_d \in \mathbf{m}(\mathbf{G})$, partitioned as in equation (4.2.7). The system $\mathbf{S}(P, C_d)$ in Figure 4.1 can be redrawn as in Figure 4.3 below, where $P = N_p D_p^{-1}$,

$C_1 = N_{c1} D_{c1}^{-1}$ and $C_2 = N_{c2} D_{c2}^{-1}$; note that $D_{c1} \xi_{c1} = e_1'$, $D_{c2} \xi_{c2} = e_2'$, $y_1' = N_{c1} \xi_{c1}$, $y_2' = N_{c2} \xi_{c2}$, where, for $i = 1, 2$, ξ_{ci} denotes the *pseudo-state* of C_i .

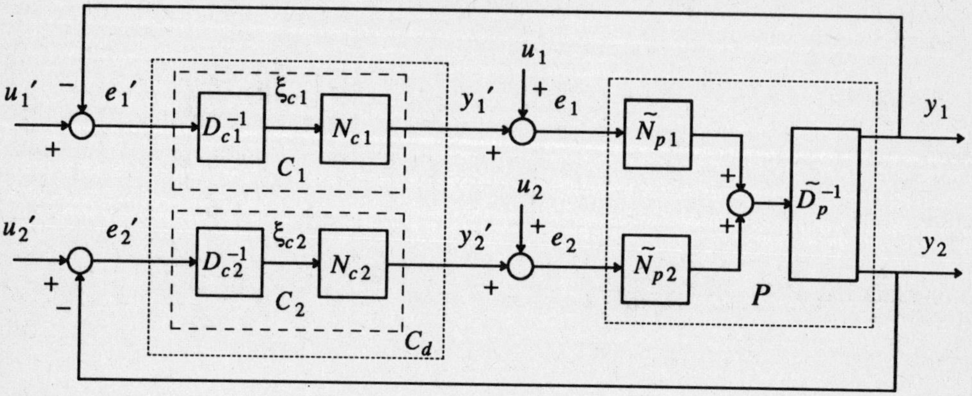

Figure 4.3. $\mathbf{S}(P, C_d)$ with $P = \tilde{D}_p^{-1} \tilde{N}_p$ and $C_d = N_c D_c^{-1}$.

The system $\mathbf{S}(P, C_d)$ is then described by equations (4.2.13)-(4.2.14) below:

$$\begin{bmatrix} \tilde{D}_{p1} D_{c1} + \tilde{N}_{p1} N_{c1} & \tilde{D}_{p2} D_{c2} + \tilde{N}_{p2} N_{c2} \end{bmatrix} \begin{bmatrix} \xi_{c1} \\ \xi_{c2} \end{bmatrix}$$

$$= \begin{bmatrix} -\tilde{N}_{p1} & -\tilde{N}_{p2} & \tilde{D}_{p1} & \tilde{D}_{p2} \end{bmatrix} \begin{bmatrix} u_1 \\ u_2 \\ u_1' \\ u_2' \end{bmatrix}, \tag{4.2.13}$$

$$\begin{bmatrix} -D_{c1} & 0 \\ 0 & -D_{c2} \\ N_{c1} & 0 \\ 0 & N_{c2} \end{bmatrix} \begin{bmatrix} \xi_{c1} \\ \xi_{c2} \end{bmatrix} = \begin{bmatrix} y_1 \\ y_2 \\ y_1' \\ y_2' \end{bmatrix} - \begin{bmatrix} 0 & 0 & I_{n_{o1}} & 0 \\ 0 & 0 & 0 & I_{n_{o2}} \\ 0 & 0 & 0 & 0 \\ 0 & 0 & 0 & 0 \end{bmatrix} \begin{bmatrix} u_1 \\ u_2 \\ u_1' \\ u_2' \end{bmatrix}. \tag{4.2.14}$$

Equations (4.2.13)-(4.2.14) are of the form

$$D_{H2}\,\xi_c = N_{HL2}\,\bar{u}$$

$$N_{HR2}\,\xi_c = \bar{y} - G_{H2}\,\bar{u}\ .$$

As in Analysis 4.2.4 (i) above, it can be easily verified that $(N_{HR2}, D_{H2}, N_{HL2})$ is a b.c. triple and that $H_{\bar{y}\bar{u}} = N_{HR2} D_{H2}^{-1} N_{HL2} + G_{H2} \in \mathfrak{m}(\mathrm{G})$ if and only if $\det D_{H2} \in \mathrm{I}$. Again by Assumption 4.2.1 (iii), $H_{\bar{y}\bar{u}} \in \mathfrak{m}(\mathrm{G})$ and hence, $\det D_{H2} \in \mathrm{I}$. Consequently, $(N_{HR2}, D_{H2}, N_{HL2}, G_{H2})$ is a b.c.f.r. of $H_{\bar{y}\bar{u}}$ and hence, $\det D_{H2}$ is a characteristic determinant of $H_{\bar{y}\bar{u}}$.

(iii) Analysis of $\mathbf{S}(P, C_d)$ with $P = N_{pr} D^{-1} N_{pl} + G$ and $C_d = \tilde{D}_c^{-1}\tilde{N}_c$

Let (N_{pr}, D, N_{pl}, G) be any b.c.f.r. of $P \in \mathfrak{m}(\mathrm{G})$, partitioned as in equation (4.2.3) and let $(\tilde{D}_c, \tilde{N}_c)$ be any l.c.f.r. of $C_d \in \mathfrak{m}(\mathrm{G})$, partitioned as in equation (4.2.6). The system $\mathbf{S}(P, C_d)$ in Figure 4.1 can be redrawn as in Figure 4.4 below, where $P = N_{pr} D^{-1} N_{pl} + G$ and $C_d = \tilde{D}_c^{-1}\tilde{N}_c$; note that $D\,\xi_x = N_{pl1}\,e_1 + N_{pl2}\,e_2$, $y_1 = N_{pr1}\,\xi_x + G_{11}\,e_1 + G_{12}\,e_2$, $y_2 = N_{pr2}\,\xi_x + G_{21}\,e_1 + G_{22}\,e_2$, where ξ_x denotes the *pseudo-state* of P.

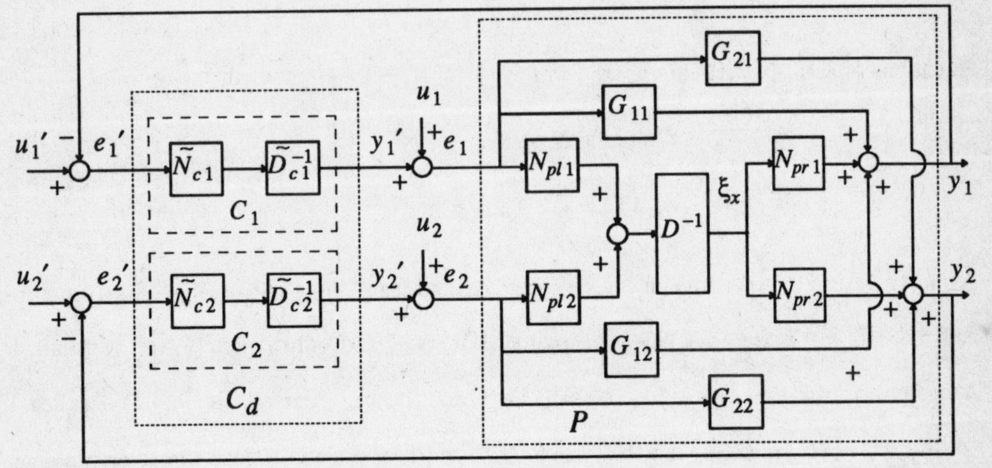

Figure 4.4. $\mathbf{S}(P, C_d)$ with $P = N_{pr} D^{-1} N_{pl} + G$ and $C_d = \tilde{D}_c^{-1}\tilde{N}_c$.

The system $\mathbf{S}(P, C_d)$ is then described by equations (4.2.15)-(4.2.16):

$$\begin{bmatrix} D & -N_{pl1} & -N_{pl2} \\ \tilde{N}_{c1} N_{pr1} & \tilde{D}_{c1} + \tilde{N}_{c1} G_{11} & \tilde{N}_{c1} G_{12} \\ \tilde{N}_{c2} N_{pr2} & \tilde{N}_{c2} G_{21} & \tilde{D}_{c2} + \tilde{N}_{c2} G_{22} \end{bmatrix} \begin{bmatrix} \xi_x \\ y_1' \\ y_2' \end{bmatrix}$$

$$= \begin{bmatrix} N_{pl1} & N_{pl2} & 0 & 0 \\ -\tilde{N}_{c1} G_{11} & -\tilde{N}_{c1} G_{12} & \tilde{N}_{c1} & 0 \\ -\tilde{N}_{c2} G_{21} & -\tilde{N}_{c2} G_{22} & 0 & \tilde{N}_{c2} \end{bmatrix} \begin{bmatrix} u_1 \\ u_2 \\ u_1' \\ u_2' \end{bmatrix}, \tag{4.2.15}$$

$$\begin{bmatrix} N_{pr1} & G_{11} & G_{12} \\ N_{pr2} & G_{21} & G_{22} \\ 0 & I_{n_{i1}} & 0 \\ 0 & 0 & I_{n_{i2}} \end{bmatrix} \begin{bmatrix} \xi_x \\ y_1' \\ y_2' \end{bmatrix} = \begin{bmatrix} y_1 \\ y_2 \\ y_1' \\ y_2' \end{bmatrix} - \begin{bmatrix} G_{11} & G_{12} & 0 & 0 \\ G_{21} & G_{22} & 0 & 0 \\ 0 & 0 & 0 & 0 \\ 0 & 0 & 0 & 0 \end{bmatrix} \begin{bmatrix} u_1 \\ u_2 \\ u_1' \\ u_2' \end{bmatrix}. \tag{4.2.16}$$

Equations (4.2.15)-(4.2.16) are of the form

$$D_{H3}\, \xi_3 = N_{HL3}\, \bar{u}$$

$$N_{HR3}\, \xi_3 = \bar{y} - G_{H3}\, \bar{u} \; .$$

As in Analysis 4.2.4 (i) above, it can be easily verified that $(N_{HR3}, D_{H3}, N_{HL3})$ is a b.c. triple and that $H_{\bar{y}\bar{u}} = N_{HR3} D_{H3}^{-1} N_{HL3} + G_{H3} \in \mathbf{m}(\mathbf{G})$ if and only if $\det D_{H3} \in \mathbf{I}$. Again by Assumption 4.2.1 (iii), $H_{\bar{y}\bar{u}} \in \mathbf{m}(\mathbf{G})$ and hence, $\det D_{H3} \in \mathbf{I}$. Consequently, $(N_{HR3}, D_{H3}, N_{HL3}, G_{H3})$ is a b.c.f.r. of $H_{\bar{y}\bar{u}}$ and hence, $\det D_{H3}$ is a characteristic determinant of $H_{\bar{y}\bar{u}}$.

(iv) Analysis of $\mathbf{S}(P,C_d)$ with $P = N_{pr}\,D^{-1}N_{pl} + G$ and $C_d = N_c\,D_c^{-1}$

Let (N_{pr}, D, N_{pl}, G) be any b.c.f.r. of $P \in \mathrm{m}(\mathrm{G})$, partitioned as in equation (4.2.3) and let $(\tilde{D}_c, \tilde{N}_c)$ be any r.c.f.r. of $C_d \in \mathrm{m}(\mathrm{G})$, partitioned as in equation (4.2.7). The system $\mathbf{S}(P,C_d)$ in Figure 4.1 can be redrawn as in Figure 4.5 below, where $P = N_{pr}\,D^{-1}N_{pl} + G$ and $C_d = \tilde{D}_c^{-1}\tilde{N}_c$; note that $D\,\xi_x = N_{pl1}\,e_1 + N_{pl2}\,e_2$, $y_1 = N_{pr1}\,\xi_x + G_{11}\,e_1 + G_{12}\,e_2$, $y_2 = N_{pr2}\,\xi_x + G_{21}\,e_1 + G_{22}\,e_2$, $D_{c1}\,\xi_{c1} = e_1'$, $D_{c2}\,\xi_{c2} = e_2'$, $y_1' = N_{c1}\,\xi_{c1}$, $y_2' = N_{c2}\,\xi_{c2}$, where ξ_x denotes the *pseudo-state* of P and for $i = 1, 2$, ξ_{ci} denotes the *pseudo-state* of C_i .

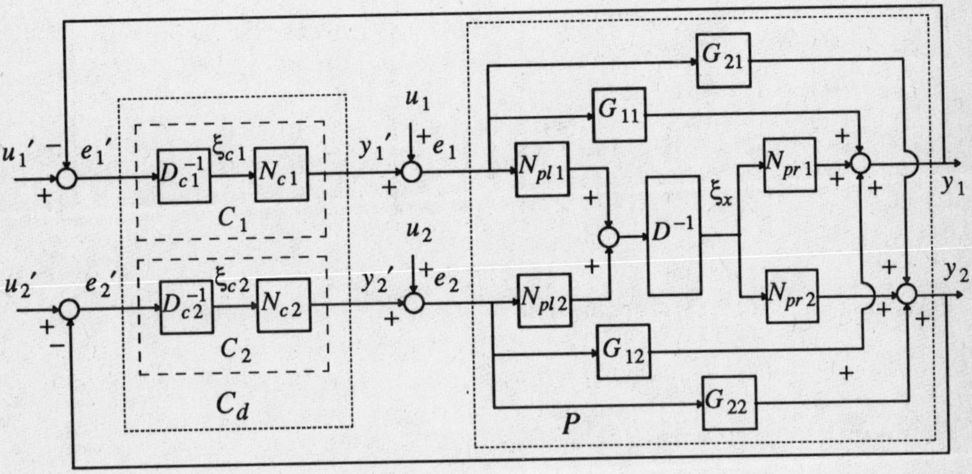

Figure 4.5. $\mathbf{S}(P,C_d)$ with $P = N_{pr}\,D^{-1}N_{pl} + G$ and $C_d = N_c\,D_c^{-1}$.

The system $\mathbf{S}(P,C_d)$ is then described by equations (4.2.17)-(4.2.18):

$$\begin{bmatrix} D & -N_{pl1}N_{c1} & -N_{pl2}N_{c2} \\ N_{pr1} & D_{c1}+G_{11}N_{c1} & G_{12}N_{c2} \\ N_{pr2} & G_{21}N_{c1} & D_{c2}+G_{22}N_{c2} \end{bmatrix} \begin{bmatrix} \xi_x \\ \xi_{c1} \\ \xi_{c2} \end{bmatrix}$$

$$= \begin{bmatrix} N_{pl1} & N_{pl2} & 0 & 0 \\ -G_{11} & -G_{12} & I_{n_{o1}} & 0 \\ -G_{21} & -G_{22} & 0 & I_{n_{o2}} \end{bmatrix} \begin{bmatrix} u_1 \\ u_2 \\ u_1' \\ u_2' \end{bmatrix}, \tag{4.2.17}$$

$$\begin{bmatrix} N_{pr1} & G_{11}N_{c1} & G_{12}N_{c2} \\ N_{pr2} & G_{21}N_{c1} & G_{22}N_{c2} \\ 0 & N_{c1} & 0 \\ 0 & 0 & N_{c2} \end{bmatrix} \begin{bmatrix} \xi_x \\ \xi_{c1} \\ \xi_{c2} \end{bmatrix} = \begin{bmatrix} y_1 \\ y_2 \\ y_1' \\ y_2' \end{bmatrix} - \begin{bmatrix} G_{11} & G_{12} & 0 & 0 \\ G_{21} & G_{22} & 0 & 0 \\ 0 & 0 & 0 & 0 \\ 0 & 0 & 0 & 0 \end{bmatrix} \begin{bmatrix} u_1 \\ u_2 \\ u_1' \\ u_2' \end{bmatrix}. \tag{4.2.18}$$

Equations (4.2.17)-(4.2.18) are of the form

$$D_{H4}\,\xi_4 = N_{HL4}\,\bar{u}$$

$$N_{HR4}\,\xi_4 = \bar{y} - G_{H4}\,\bar{u} .$$

As in Analysis 4.2.4 (i) above, it can be easily verified that $(N_{HR4}, D_{H4}, N_{HL4})$ is a b.c. triple and that $H_{\bar{y}\bar{u}} = N_{HR4}D_{H4}^{-1}N_{HL4} + G_{H4} \in \mathrm{m}(\mathrm{G})$ if and only if $\det D_{H4} \in \mathrm{I}$. Again by Assumption 4.2.1 (iii), $H_{\bar{y}\bar{u}} \in \mathrm{m}(\mathrm{G})$ and hence, $\det D_{H4} \in \mathrm{I}$. Consequently, $(N_{HR4}, D_{H4}, N_{HL4}, G_{H4})$ is a b.c.f.r. of $H_{\bar{y}\bar{u}}$ and hence, $\det D_{H4}$ is a characteristic determinant of $H_{\bar{y}\bar{u}}$. □

Theorem 4.2.5. (**H–stability of** $\mathrm{S}(P, C_d)$)

Let Assumptions 4.2.1 (i) and (ii) hold; let (N_p, D_p) be any r.c.f.r., $(\tilde{D}_p, \tilde{N}_p)$ be any l.c.f.r., (N_{pr}, D, N_{pl}, G) be any b.c.f.r. over $\mathrm{m}(\mathrm{H})$ of $P \in \mathrm{m}(\mathrm{G})$, partitioned as in equations

(4.2.1), (4.2.2) and (4.2.3), respectively; let $(\tilde{D}_c, \tilde{N}_c)$ be any l.c.f.r., (N_c, D_c) be any r.c.f.r. over $\mathrm{m}(\mathrm{H})$ of $C \in \mathrm{m}(\mathrm{G})$, partitioned as in equations (4.2.6) and (4.2.7), respectively. Under these assumptions, the following five statements are equivalent:

(i) $\mathrm{S}(P, C_d)$ is H–stable ;

(ii) $$D_{H1} := \left[\tilde{D}_c D_p + \tilde{N}_c N_p \right] = \begin{bmatrix} \tilde{D}_{c1} D_{p1} + \tilde{N}_{c1} N_{p1} \\ \tilde{D}_{c2} D_{p2} + \tilde{N}_{c2} N_{p2} \end{bmatrix}$$ is H–unimodular ; (4.2.19)

(iii) $$D_{H2} := \left[\tilde{D}_p D_c + \tilde{N}_p N_c \right]$$

$$= \left[\tilde{D}_{p1} D_{c1} + \tilde{N}_{p1} N_{c1} \quad \tilde{D}_{p2} D_{c2} + \tilde{N}_{p2} N_{c2} \right]$$ is H–unimodular ; (4.2.20)

(iv) $$D_{H3} := \begin{bmatrix} D & -N_{pl} \\ \tilde{N}_c N_{pr} & \tilde{D}_c + \tilde{N}_c G \end{bmatrix}$$

$$= \begin{bmatrix} D & -N_{pl1} & -N_{pl2} \\ \tilde{N}_{c1} N_{pr1} & \tilde{D}_{c1} + \tilde{N}_{c1} G_{11} & \tilde{N}_{c2} G_{21} \\ \tilde{N}_{c2} N_{pr2} & \tilde{N}_{c1} G_{21} & \tilde{D}_{c2} + \tilde{N}_{c2} G_{22} \end{bmatrix}$$ is H–unimodular ; (4.2.21)

(v) $$D_{H4} := \begin{bmatrix} D & -N_{pl} N_c \\ N_{pr} & D_c + G N_c \end{bmatrix}$$

$$= \begin{bmatrix} D & -N_{pl1} N_{c1} & -N_{pl2} N_{c2} \\ N_{pr1} & D_{c1} + G_{11} N_{c1} & G_{12} N_{c2} \\ N_{pr2} & G_{21} N_{c1} & D_{c2} + G_{22} N_{c2} \end{bmatrix}$$ is H–unimodular . □ (4.2.22)

Note that each of statements (i) through (v) of Theorem 4.2.5 implies that the system $S(P, C_d)$ is well-posed; consequently, we do not need to state a well-posedness assumption in the beginning of Theorem 4.2.5.

Proof

Follows from the system descriptions in Analysis 4.2.4 as in Theorem 3.2.7.

Remark 4.2.6

The denominator matrix D_{H1} in equation (4.2.19) can also be written as follows:

$$D_{H1} = \begin{bmatrix} \tilde{D}_{c1} & 0 & \tilde{N}_{c1} & 0 \\ 0 & \tilde{D}_{c2} & 0 & \tilde{N}_{c2} \end{bmatrix} \begin{bmatrix} D_{p1} \\ D_{p2} \\ N_{p1} \\ N_{p2} \end{bmatrix} = \begin{bmatrix} \tilde{D}_{c1} & \tilde{N}_{c1} & 0 & 0 \\ 0 & 0 & \tilde{D}_{c2} & \tilde{N}_{c2} \end{bmatrix} \begin{bmatrix} D_{p1} \\ N_{p1} \\ D_{p2} \\ N_{p2} \end{bmatrix} ; \tag{4.2.23}$$

and $\det D_{H1}$ can also be written as $\det D_{H1} = \det \tilde{D}_c \det(I + C_d P) \det D_p$. By normalization and due to the block-diagonal compensator structure, $D_{H1} \in \mathrm{m}(\mathbf{H})$ is H–unimodular if and only if there are *block-diagonal* matrices $V_p := \tilde{D}_c$, $U_p := \tilde{N}_c \in \mathrm{m}(\mathbf{H})$ such that

$$V_p D_p + U_p N_p = I_{n_i} . \tag{4.2.24}$$

Equation (4.2.24) is a Bezout identity where V_p , $U_p \in \mathrm{m}(\mathbf{H})$ are restricted to be *block-diagonal* as shown in equation (4.2.23).

The denominator matrix D_{H2} in equation (4.2.20) can also be written as

$$D_{H2} = \begin{bmatrix} -\tilde{N}_{p1} & -\tilde{N}_{p2} & \tilde{D}_{p1} & \tilde{D}_{p2} \end{bmatrix} \begin{bmatrix} -N_{c1} & 0 \\ 0 & -N_{c2} \\ D_{c1} & 0 \\ 0 & D_{c2} \end{bmatrix}$$

$$= \begin{bmatrix} -\tilde{N}_{p1} & \tilde{D}_{p1} & -\tilde{N}_{p2} & \tilde{D}_{p2} \end{bmatrix} \begin{bmatrix} -N_{c1} & 0 \\ D_{c1} & 0 \\ 0 & -N_{c2} \\ 0 & D_{c2} \end{bmatrix} . \tag{4.2.25}$$

Following Remark 3.2.10, the denominator matrix D_{H3} in equation (3.2.21) is H–unimodular if and only if $\tilde{D}_c Y + \tilde{N}_c (N_{pr} X + G Y)$ is H–unimodular, where

$$\tilde{D}_c Y + \tilde{N}_c (N_{pr} X + G Y)$$

$$= \begin{bmatrix} \tilde{D}_{c1} & 0 \\ 0 & \tilde{D}_{c2} \end{bmatrix} \begin{bmatrix} Y_1 \\ Y_2 \end{bmatrix} + \begin{bmatrix} \tilde{N}_{c1} & 0 \\ 0 & \tilde{N}_{c2} \end{bmatrix} \begin{bmatrix} N_{pr1} X + G_{11} Y_1 + G_{12} Y_2 \\ N_{pr2} X + G_{21} Y_1 + G_{22} Y_2 \end{bmatrix}$$

$$= \begin{bmatrix} \tilde{D}_{c1} & \tilde{N}_{c1} & 0 & 0 \\ 0 & 0 & \tilde{D}_{c2} & \tilde{N}_{c2} \end{bmatrix} \begin{bmatrix} Y_1 \\ N_{pr1} X + G_{11} Y_1 + G_{12} Y_2 \\ Y_2 \\ N_{pr2} X + G_{21} Y_1 + G_{22} Y_2 \end{bmatrix} . \qquad (4.2.26)$$

Similarly, the denominator matrix D_{H4} in equation (3.2.22) is H–unimodular if and only if $(\tilde{X} N_{pl} + \tilde{Y} G) N_c + \tilde{Y} D_c$ is H–unimodular, where

$$(\tilde{X} N_{pl} + \tilde{Y} G) N_c + \tilde{Y} D_c$$

$$= \begin{bmatrix} \tilde{X} N_{pl1} + \tilde{Y}_1 G_{11} + \tilde{Y}_2 G_{21} & \tilde{X} N_{pl2} + \tilde{Y}_1 G_{12} + \tilde{Y}_2 G_{22} \end{bmatrix} \begin{bmatrix} N_{c1} & 0 \\ 0 & N_{c2} \end{bmatrix}$$

$$+ \begin{bmatrix} \tilde{Y}_1 & \tilde{Y}_2 \end{bmatrix} \begin{bmatrix} D_{c1} & 0 \\ 0 & D_{c2} \end{bmatrix} =$$

$$\begin{bmatrix} -(\tilde{X} N_{pl1} + \tilde{Y}_1 G_{11} + \tilde{Y}_2 G_{21}) & \tilde{Y}_1 & -(\tilde{X} N_{pl2} + \tilde{Y}_1 G_{12} + \tilde{Y}_2 G_{22}) & \tilde{Y}_2 \end{bmatrix} \begin{bmatrix} -N_{c1} & 0 \\ D_{c1} & 0 \\ 0 & -N_{c2} \\ 0 & D_{c2} \end{bmatrix} .$$

(4.2.27)

4.3 TWO-CHANNEL DECENTRALIZED FEEDBACK COMPENSATORS

Throughout this section, we assume that the plant P satisfies Assumption 4.2.1 (i).

Definition 4.3.1. (Decentralized H–stabilizing compensator C_d)

(i) C_d is called a *decentralized* H–*stabilizing compensator for* P (abbreviated as: C_d H–*stabilizes* P) iff $C_d = \begin{bmatrix} C_1 & 0 \\ 0 & C_2 \end{bmatrix} \in \mathbf{G}^{n_i \times n_o}$ and the system $\mathbf{S}(P, C_d)$ is H–stable.

(ii) The set

$$S_d(P) := \{ \; C_d = \begin{bmatrix} C_1 & 0 \\ 0 & C_2 \end{bmatrix} \; | \; C_d \text{ H–}stabilizes\; P \; \}$$

is called the *set of all decentralized* H–*stabilizing compensators for* P in the system $\mathbf{S}(P, C_d)$.

Comment 4.3.2

In Chapter 3 (Theorem 3.2.11) we showed that the set $S(P)$ of all (full-feedback) compensators that H–stabilize P in the system $\mathbf{S}(P, C)$ is given by

$$S(P) = \{ \, (V_p - Q\tilde{N}_p)^{-1} (U_p + Q\tilde{D}_p) \; | \; Q \in \mathbf{H}^{n_i \times n_o}, \det(V_p - Q\tilde{N}_p) \in \mathbf{I} \, \} ; \tag{4.3.1}$$

equivalently,

$$S(P) = \{ \, (\tilde{U}_p + D_p Q)(\tilde{V}_p - N_p Q)^{-1} \; | \; Q \in \mathbf{H}^{n_i \times n_o}, \det(\tilde{V}_p - N_p Q) \in \mathbf{I} \, \} , \tag{4.3.2}$$

where (N_p, D_p) is any r.c.f.r. and $(\tilde{D}_p, \tilde{N}_p)$ is any l.c.f.r. of $P \in \mathbf{G}^{n_o \times n_i}$ and V_p, U_p, $\tilde{V}_p$, $\tilde{U}_p \in \mathfrak{m}(\mathbf{H})$ are as in the generalized Bezout identity (2.3.12).

The class of all *decentralized* H–stabilizing compensators $S_d(P)$ is more complicated. (Note that $S_d(P)$ is a subset of $S(P)$). For the existence of such *decentralized* compensators that H–stabilize P in the system $\mathbf{S}(P, C_d)$, the plant P must satisfy *additional* conditions which are not required for the existence of *full-feedback* compensators that would H–stabilize P in the system $\mathbf{S}(P, C)$; these conditions are due to the block-diagonal structure of the compensator. □

Theorem 4.3.3 below establishes the necessary and sufficient conditions on P for the existence of decentralized H–stabilizing feedback compensators for P :

Theorem 4.3.3. (Conditions on $P = N_p D_p^{-1} = \tilde{D}_p^{-1} \tilde{N}_p$ for decentralized H–stabilizability) Let P satisfy Assumption 4.2.1 (i); furthermore, let $P \in \mathfrak{m}(\mathbf{G}_s)$; then the following three conditions are equivalent:

(i) There exists a decentralized H–stabilizing compensator C_d for P ;

(ii) Any r.c.f.r. (N_p, D_p) of P, partitioned as in equation (4.2.1), satisfies conditions (4.3.3) and (4.3.4) below:

$$E_1 \begin{bmatrix} D_{p1} \\ N_{p1} \end{bmatrix} R = \begin{bmatrix} I_{ni1} & 0 \\ 0 & W_{12} \end{bmatrix}, \tag{4.3.3}$$

$$E_2 \begin{bmatrix} D_{p2} \\ N_{p2} \end{bmatrix} R = \begin{bmatrix} 0 & I_{ni2} \\ W_{21} & 0 \end{bmatrix}, \tag{4.3.4}$$

(iii) Any l.c.f.r. $(\tilde{D}_p, \tilde{N}_p)$ of P, partitioned as in equation (4.2.2), satisfies conditions (4.3.5) and (4.3.6) below:

$$L \begin{bmatrix} -\tilde{N}_{p1} & \tilde{D}_{p1} \end{bmatrix} E_1^{-1} = \begin{bmatrix} 0 & I_{no1} \\ -W_{21} & 0 \end{bmatrix}, \tag{4.3.5}$$

$$L \begin{bmatrix} -\tilde{N}_{p2} & \tilde{D}_{p2} \end{bmatrix} E_2^{-1} = \begin{bmatrix} -W_{12} & 0 \\ 0 & I_{n_{o2}} \end{bmatrix}, \tag{4.3.6}$$

where $E_1 \in H^{(n_{i1}+n_{o1}) \times (n_{i1}+n_{o1})}$ is H–unimodular, $E_2 \in H^{(n_{i2}+n_{o2}) \times (n_{i2}+n_{o2})}$ is H–unimodular, $R \in H^{n_i \times n_i}$ is H–unimodular and $L \in H^{n_o \times n_o}$ is H–unimodular; the matrices $W_{12} \in H^{n_{o1} \times n_{i2}}$ and $W_{21} \in H^{n_{o2} \times n_{i1}}$ are H–stable in equations (4.3.3) through (4.3.6).

Comment 4.3.4

(i) Statements (ii) and (iii) of Theorem 4.3.3 are equivalent for all $P \in m(G)$. Statement (i) implies statement (ii) for all $P \in m(G)$; the assumption that $P \in m(G_s)$ is only needed in proving that statement (ii) (equivalently, statement (iii)) implies statement (i).

(ii) Since $\tilde{D}_p N_p = \tilde{N}_p D_p$, the H–unimodular matrices $E_1 \in H^{(n_{i1}+n_{o1}) \times (n_{i1}+n_{o1})}$, $E_2 \in H^{(n_{i2}+n_{o2}) \times (n_{i2}+n_{o2})}$ and the H–stable matrices $W_{12} \in H^{n_{o1} \times n_{i2}}$, $W_{21} \in H^{n_{o2} \times n_{i1}}$ in equations (4.3.3)-(4.3.4) are the same as the ones in equations (4.3.5)-(4.3.6).

(iii) Let (N_p, D_p) be an r.c.f.r. of P; then by Lemma 2.3.4 (i), $(X_p, Y_p) := (N_p R, D_p R)$ is also an r.c.f.r. of P, where the matrix $R \in H^{n_i \times n_i}$ is H–unimodular. Let $X_p := \begin{bmatrix} X_{p1} \\ X_{p2} \end{bmatrix}$, $Y_p := \begin{bmatrix} Y_{p1} \\ Y_{p2} \end{bmatrix}$. By Theorem 4.3.3, P can be H–stabilized by a decentralized compensator C_d if and only if some r.c.f.r. (X_p, Y_p) of P is of the form

$$\begin{bmatrix} Y_{p1} \\ X_{p1} \\ \cdots \\ Y_{p2} \\ X_{p2} \end{bmatrix} = \begin{bmatrix} D_{p1} \\ N_{p1} \\ \cdots \\ D_{p2} \\ N_{p2} \end{bmatrix} R = \begin{bmatrix} E_1^{-1} & 0 \\ \cdots & \cdots \\ 0 & E_2^{-1} \end{bmatrix} \begin{bmatrix} I_{n_{i1}} & 0 \\ 0 & W_{12} \\ \cdots & \cdots \\ 0 & I_{n_{i2}} \\ W_{21} & 0 \end{bmatrix}, \tag{4.3.7}$$

where $E_1, E_2 \in \mathfrak{m}(\mathbf{H})$ are H–unimodular and $W_{12}, W_{21} \in \mathfrak{m}(\mathbf{H})$.

Similarly, let $(\tilde{D}_p, \tilde{N}_p)$ be an l.c.f.r. of P ; then $(\tilde{Y}_p, \tilde{X}_p) := (L\tilde{D}_p, L\tilde{N}_p)$ is also an l.c.f.r. of P , where the matrix $L \in \mathbf{H}^{n_o \times n_o}$ is H–unimodular. Let $\tilde{Y}_p := \begin{bmatrix} \tilde{Y}_{p1} & \tilde{Y}_{p2} \end{bmatrix}$, $\tilde{X}_p := \begin{bmatrix} \tilde{X}_{p1} & \tilde{X}_{p2} \end{bmatrix}$. By Theorem 4.3.3, P can be H–stabilized by a decentralized compensator C_d if and only if some l.c.f.r. $(\tilde{Y}_p, \tilde{X}_p)$ of P is of the form

$$\begin{bmatrix} -\tilde{X}_{p1} & \tilde{Y}_{p1} & \vdots & -\tilde{X}_{p2} & \tilde{Y}_{p2} \end{bmatrix} - L\begin{bmatrix} \tilde{N}_{p1} & \tilde{D}_{p1} & \vdots & -\tilde{N}_{p2} & \tilde{D}_{p2} \end{bmatrix}$$

$$= \begin{bmatrix} 0 & I_{n_{o1}} & \vdots & -W_{12} & 0 \\ -W_{21} & 0 & \vdots & 0 & I_{n_{o2}} \end{bmatrix} \begin{bmatrix} E_1 & 0 \\ \cdots & \cdots \\ 0 & E_2 \end{bmatrix}, \tag{4.3.8}$$

where $E_1, E_2 \in \mathfrak{m}(\mathbf{H})$ are H–unimodular and $W_{12}, W_{21} \in \mathfrak{m}(\mathbf{H})$ are H–unimodular matrices of appropriate dimensions.

(iv) Suppose that the plant P is H–stable; then following Comment 2.4.7 (iii), (P, I_{n_i}) is an r.c.f.r. and (I_{n_o}, P) is an l.c.f.r. of $P = \begin{bmatrix} P_{11} & P_{12} \\ P_{21} & P_{22} \end{bmatrix}$. Clearly, conditions (4.3.3)-(4.3.4) (equivalently, conditions (4.3.5)-(4.3.6)) hold whenever $P \in \mathfrak{m}(\mathbf{H})$. In this case, $\begin{bmatrix} D_{p1} \\ N_{p1} \end{bmatrix} = \begin{bmatrix} I_{n_{i1}} & 0 \\ P_{11} & P_{12} \end{bmatrix}$ and $\begin{bmatrix} D_{p2} \\ N_{p2} \end{bmatrix} = \begin{bmatrix} 0 & I_{n_{i2}} \\ P_{21} & P_{22} \end{bmatrix}$; hence, the H–unimodular matrices in conditions (4.3.3)-(4.3.4) (equivalently, conditions (4.3.5)-(4.3.6)) can be chosen as $E_1 = \begin{bmatrix} I_{n_{i1}} & 0 \\ -P_{11} & I_{n_{o1}} \end{bmatrix}$, $E_2 = \begin{bmatrix} I_{n_{i2}} & 0 \\ -P_{22} & I_{n_{o2}} \end{bmatrix}$, $R = I_{n_i}$; then the matrices $W_{12} = P_{12}$, $W_{21} = P_{21}$.

(v) Let P satisfy Assumption 4.2.1 (i). Now suppose that $P_{12} = 0$, i.e., $P = \begin{bmatrix} P_{11} & 0 \\ P_{21} & P_{22} \end{bmatrix}$ is lower block-triangular; then $(N_p, D_p) :=$

$(\begin{bmatrix} N_{11} & 0 \\ N_{21} & N_{22} \end{bmatrix}, \begin{bmatrix} D_{11} & 0 \\ D_{21} & D_{22} \end{bmatrix})$ is an r.c.f.r. of P, where (N_{22}, D_{22}) is an r.c.f.r. of P_{22}. Since (N_{22}, D_{22}) is an r.c. pair, there is an H–unimodular matrix $E_2 \in m(H)$ such that $E_2 \begin{bmatrix} D_{22} \\ N_{22} \end{bmatrix} = \begin{bmatrix} I_{n_{i2}} \\ 0 \end{bmatrix}$. Clearly, condition (4.3.4) holds. Now condition (4.3.3) holds if and only if (N_{11}, D_{11}) is also an r.c. pair, i.e., (N_{11}, D_{11}) is an r.c.f.r. of P_{11}; by Theorem 4.3.3, in the case that $P_{12} = 0$, there exists a decentralized H–stabilizing compensator if and only if (N_{11}, D_{11}) is also an r.c. pair. If P is lower block-triangular, then $W_{12} = 0$, W_{21}

$= \begin{bmatrix} 0 & I_{n_{o2}} \end{bmatrix} E_2 \begin{bmatrix} D_{21} \\ N_{21} \end{bmatrix}$, $R = \begin{bmatrix} I_{n_{i1}} & 0 \\ -\begin{bmatrix} I_{n_{i2}} & 0 \end{bmatrix} E_2 \begin{bmatrix} D_{21} \\ N_{21} \end{bmatrix} & I_{n_{i2}} \end{bmatrix}$.

Similarly, if $P_{21} = 0$, i.e., $P = \begin{bmatrix} P_{11} & P_{12} \\ 0 & P_{22} \end{bmatrix}$ is upper block-triangular, then (N_p, D_p)

$:= (\begin{bmatrix} N_{11} & N_{12} \\ 0 & N_{22} \end{bmatrix}, \begin{bmatrix} D_{11} & D_{12} \\ 0 & D_{22} \end{bmatrix})$ is an r.c.f.r. of P, where (N_{11}, D_{11}) is an r.c.f.r. of P_{11}. This time condition (4.3.3) holds automatically. Now condition (4.3.4) holds if and only if (N_{22}, D_{22}) is also an r.c. pair, i.e., (N_{22}, D_{22}) is an r.c.f.r. of P_{22}; by Theorem 4.3.3, in the case that $P_{21} = 0$, there exists a decentralized H–stabilizing compensator if and only if (N_{22}, D_{22}) is also an r.c. pair. If P is upper block-triangular, then $W_{21} = 0$.

(vi) In Section 4.4 we show that, if H is the ring R_u as in Section 2.2, then conditions (4.3.3)-(4.3.4) (equivalently, conditions (4.3.5)-(4.3.6)) on $P \in \mathbb{R}_p(s)^{n_o \times n_i}$ are equivalent to the condition that the system $S(P, C_d)$ has *no fixed-eigenvalues* in $\bar{u}$.

(vii) Suppose that $P \in m(G_s)$ is given by a b.c.f.r. (N_{pr}, D, N_{pl}, G) and $C_d \in m(G)$ is given by an l.c.f.r. $(\tilde{D}_c, \tilde{N}_c)$ as in case (iii) of Analysis 4.2.4. Considering equation (4.2.26), apply Theorem 4.3.3 to the r.c.f.r. $(N_p, D_p) := (N_{pr} X + G Y, Y)$ of P; then equations (4.3.3)-(4.3.4) imply that $P = N_{pr} D^{-1} N_{pl} + G \in m(G_s)$ can be H–stabilized by a decentralized compensator C_d if and only if

$$E_1 \begin{bmatrix} Y_1 \\ N_{pr1}X + G_{11}Y_1 + G_{12}Y_2 \end{bmatrix} R = \begin{bmatrix} I_{ni1} & 0 \\ 0 & W_{12} \end{bmatrix} \quad \text{and} \tag{4.3.9}$$

$$E_2 \begin{bmatrix} Y_2 \\ N_{pr2}X + G_{21}Y_1 + G_{22}Y_2 \end{bmatrix} R = \begin{bmatrix} 0 & I_{ni2} \\ W_{21} & 0 \end{bmatrix}, \tag{4.3.10}$$

where $E_1 \in \mathbf{H}^{(n_{i1}+n_{o1}) \times (n_{i1}+n_{o1})}$, $E_2 \in \mathbf{H}^{(n_{i2}+n_{o2}) \times (n_{i2}+n_{o2})}$ and $R \in \mathbf{H}^{n_i \times n_i}$ are H–unimodular; $W_{12} \in \mathbf{H}^{n_{o1} \times n_{i2}}$, $W_{21} \in \mathbf{H}^{n_{o2} \times n_{i1}}$.

Similarly if P is given by a b.c.f.r. (N_{pr}, D, N_{pl}, G) and C_d is given by an r.c.f.r. (N_c, D_c) as in case **(iv)** of Analysis 4.2.4, then considering equation (4.2.27), we apply Theorem 4.3.3 to the l.c.f.r. $(\tilde{D}_p, \tilde{N}_p) := (\tilde{Y}, \tilde{X} N_{pl} + \tilde{Y} G)$ of P. Following equations (4.3.5)-(4.3.6), P can be H–stabilized by a decentralized compensator C_d if and only if

$$L \begin{bmatrix} -(\tilde{X}N_{pl1} + \tilde{Y}_1 G_{11} + \tilde{Y}_2 G_{21}) & \tilde{Y}_1 \end{bmatrix} E_1^{-1} = \begin{bmatrix} 0 & I_{no1} \\ -W_{21} & 0 \end{bmatrix} \tag{4.3.11}$$

and

$$L \begin{bmatrix} -(\tilde{X}N_{pl2} + \tilde{Y}_1 G_{12} + \tilde{Y}_2 G_{22}) & \tilde{Y}_2 \end{bmatrix} E_2^{-1} = \begin{bmatrix} -W_{12} & 0 \\ 0 & I_{no2} \end{bmatrix} \tag{4.3.12}$$

where $E_1^{-1} \in \mathbf{H}^{(n_{i1}+n_{o1}) \times (n_{i1}+n_{o1})}$, $E_2^{-1} \in \mathbf{H}^{(n_{i2}+n_{o2}) \times (n_{i2}+n_{o2})}$ and $L \in \mathbf{H}^{n_o \times n_o}$ are H–unimodular; $W_{12} \in \mathbf{H}^{n_{o1} \times n_{i2}}$ and $W_{21} \in \mathbf{H}^{n_{o2} \times n_{i1}}$.

Equations (4.3.9)-(4.3.10) (equivalently, (4.3.11)-(4.3.12)) are useful in Section 4.4 where we explain *"rank-tests"* for decentralized H–stabilizability in terms of the state-space representation of P.

Proof of Theorem 4.3.3 *Statement (i) is equivalent to statement (ii):*

Suppose that statement (ii) holds, i.e., any r.c.f.r. (N_p, D_p) of P satisfies conditions (4.3.3)-(4.3.4); then

$$\begin{bmatrix} E_1 & 0 \\ \cdots & \cdots \\ 0 & E_2 \end{bmatrix} \begin{bmatrix} D_{p1} \\ N_{p1} \\ \cdots \\ D_{p2} \\ N_{p2} \end{bmatrix} R = \begin{bmatrix} I_{n_{i1}} & 0 \\ 0 & W_{12} \\ \cdots & \cdots \\ 0 & I_{n_{i2}} \\ W_{21} & 0 \end{bmatrix} . \tag{4.3.13}$$

Refer to equation (4.2.23) and consider the decentralized compensator $C_d = \begin{bmatrix} C_1 & 0 \\ 0 & C_2 \end{bmatrix} =$ $\begin{bmatrix} \tilde{D}_{c1}^{-1}\tilde{N}_{c1} & 0 \\ 0 & \tilde{D}_{c2}^{-1}\tilde{N}_{c2} \end{bmatrix}$, where $\tilde{D}_{c1}, \tilde{D}_{c2}, \tilde{N}_{c1}, \tilde{N}_{c2}$ are given by

$$\begin{bmatrix} \tilde{D}_{c1} & \tilde{N}_{c1} \end{bmatrix} = \begin{bmatrix} I_{n_{i1}} & 0 \end{bmatrix} E_1 , \tag{4.3.14}$$

$$\begin{bmatrix} \tilde{D}_{c2} & \tilde{N}_{c2} \end{bmatrix} = \begin{bmatrix} I_{n_{i2}} & 0 \end{bmatrix} E_2 . \tag{4.3.15}$$

Since $E_1, E_2 \in \mathrm{m}(\mathrm{H})$, clearly $\tilde{D}_{c1}, \tilde{D}_{c2}, \tilde{N}_{c1}, \tilde{N}_{c2} \in \mathrm{m}(\mathrm{H})$. Since for $k = 1, 2$, E_k is H–unimodular, $(\tilde{D}_{ck}, \tilde{N}_{ck})$ is an l.c. pair.

With $(\tilde{D}_{c1}, \tilde{N}_{c1})$ as in equation (4.3.14) and $(\tilde{D}_{c2}, \tilde{N}_{c2})$ as in equation (4.3.15) and (N_p, D_p) as in equation (4.3.13), the denominator matrix D_{H1} in equation (4.2.23) is H–unimodular since

$$D_{H1} = \begin{bmatrix} \tilde{D}_{c1} & 0 & \tilde{N}_{c1} & 0 \\ 0 & \tilde{D}_{c2} & 0 & \tilde{N}_{c2} \end{bmatrix} \begin{bmatrix} D_{p1} \\ D_{p2} \\ N_{p1} \\ N_{p2} \end{bmatrix}$$

$$= \begin{bmatrix} \tilde{D}_{c1} & \tilde{N}_{c1} & 0 & 0 \\ 0 & 0 & \tilde{D}_{c2} & \tilde{N}_{c2} \end{bmatrix} \begin{bmatrix} D_{p1} \\ N_{p1} \\ D_{p2} \\ N_{p2} \end{bmatrix} = \begin{bmatrix} I_{n_{i1}} & 0 \\ 0 & I_{n_{i2}} \end{bmatrix} R^{-1} ; \qquad (4.3.16)$$

note that $R \in \mathbf{H}^{n_i \times n_i}$ is H–unimodular. Equation (4.3.16) implies that

$$\tilde{D}_c D_p + \tilde{N}_c N_p = R^{-1} , \qquad (4.3.17)$$

where $R \in \mathbf{H}^{n_i \times n_i}$ is H–unimodular. From equation (4.3.17),

$$\det\tilde{D}_c \det D_p = \det(I_{n_i} - \tilde{N}_c N_p R) \det R^{-1} . \qquad (4.3.18)$$

Now $P \in \mathfrak{m}(\mathbf{G}_s)$ implies that $N_p \in \mathfrak{m}(\mathbf{G}_s)$ and therefore, $\tilde{N}_c N_p R \in \mathfrak{m}(\mathbf{G}_s)$; hence, $\det(I_{n_i} - \tilde{N}_c N_p R) \in \mathbf{I}$. But since $\det R \in \mathbf{J}$, by equation (4.3.18), $\det\tilde{D}_c \det D_p \in \mathbf{I}$ and hence, by Lemma 2.3.3 (ii), $\det D_p \in \mathbf{I}$ and $\det\tilde{D}_c \in \mathbf{I}$; but $\det\tilde{D}_c = \det \begin{bmatrix} \tilde{D}_{c1} & 0 \\ 0 & \tilde{D}_{c2} \end{bmatrix} \in \mathbf{I}$ if and only if $\det\tilde{D}_{c1} \in \mathbf{I}$ and $\det\tilde{D}_{c2} \in \mathbf{I}$, where, from equations (4.3.14)-(4.3.15),

$$\tilde{D}_{c1} = \begin{bmatrix} I_{n_{i1}} & 0 \end{bmatrix} E_1 \begin{bmatrix} I_{n_{i1}} \\ 0 \end{bmatrix} , \quad \tilde{D}_{c2} = \begin{bmatrix} I_{n_{i2}} & 0 \end{bmatrix} E_2 \begin{bmatrix} I_{n_{i2}} \\ 0 \end{bmatrix} . \qquad (4.3.19)$$

This proves that $(\tilde{D}_{c1}, \tilde{N}_{c1})$ given by equation (4.3.14) is an l.c.f.r. of $C_1 \in \mathfrak{m}(\mathbf{G})$ and $(\tilde{D}_{c2}, \tilde{N}_{c2})$ given by equation (4.3.15) is a l.c.f.r. of $C_2 \in \mathfrak{m}(\mathbf{G})$. Now since equation (4.3.16) implies that D_{H1} is H–unimodular, the system $\mathbf{S}(P, C_d)$ is H–stable with this choice of decentralized compensator $C_d = \begin{bmatrix} C_1 & 0 \\ 0 & C_2 \end{bmatrix} = \begin{bmatrix} \tilde{D}_{c1}^{-1}\tilde{N}_{c1} & 0 \\ 0 & \tilde{D}_{c2}^{-1}\tilde{N}_{c2} \end{bmatrix}$.

Now suppose that statement (i) holds. Let $C_d = \begin{bmatrix} C_1 & 0 \\ 0 & C_2 \end{bmatrix}$ be a decentralized H–stabilizing compensator for $P \in \mathfrak{m}(\mathbf{G}_s)$; by Definition 4.3.1, C_d satisfies Assumption

4.2.1 (ii) and the system $S(P, C_d)$ is H–stable. Let $(\tilde{D}_{c1}, \tilde{N}_{c1})$ be an l.c.f.r. of C_1 and $(\tilde{D}_{c2}, \tilde{N}_{c2})$ be an l.c.f.r. of C_2. Let (N_p, D_p) be any r.c.f.r. of P, partitioned as in equation (4.2.1). By Theorem 4.2.5, since $S(P, C_d)$ is H–stable, the matrix D_{H1} given by equation (4.2.19) is H–unimodular; hence, $D_{H1}^{-1} \in \mathrm{m}(\mathrm{H})$. Let $D_{H1}^{-1} =: R$ and let

$$\begin{bmatrix} D_{p1} \\ N_{p1} \\ D_{p2} \\ N_{p2} \end{bmatrix} R =: \begin{bmatrix} D_{11} & D_{12} \\ N_{11} & N_{12} \\ D_{21} & D_{22} \\ N_{21} & N_{22} \end{bmatrix} . \tag{4.3.20}$$

By Lemma 2.3.4 (i), since R is H–unimodular, $(N_p R, D_p R)$ is also an r.c.f.r. of P. By equations (4.2.19) and (4.3.20),

$$\begin{bmatrix} \tilde{D}_{c1} & \tilde{N}_{c1} & 0 & 0 \\ 0 & 0 & \tilde{D}_{c2} & \tilde{N}_{c2} \end{bmatrix} \begin{bmatrix} D_{p1} \\ N_{p1} \\ D_{p2} \\ N_{p2} \end{bmatrix}$$

$$= \begin{bmatrix} \tilde{D}_{c1} & \tilde{N}_{c1} & 0 & 0 \\ 0 & 0 & \tilde{D}_{c2} & \tilde{N}_{c2} \end{bmatrix} \begin{bmatrix} D_{11} & D_{12} \\ N_{11} & N_{12} \\ D_{21} & D_{22} \\ N_{21} & N_{22} \end{bmatrix} = \begin{bmatrix} I_{n_{i1}} & 0 \\ 0 & I_{n_{i2}} \end{bmatrix} . \tag{4.3.21}$$

Let (N_{c1}, D_{c1}) be an r.c.f.r. of C_1 and (N_{c2}, D_{c2}) be an r.c.f.r. of C_2; then $-\tilde{D}_{c1} N_{c1} + \tilde{N}_{c1} D_{c1} = 0$ and $-\tilde{D}_{c2} N_{c2} + \tilde{N}_{c2} D_{c2} = 0$. By equation (4.3.21), it follows from Lemma 2.3.7 that there are matrices V_{c1}, U_{c1} V_{c2}, $U_{c2} \in \mathrm{m}(\mathrm{H})$, where $V_{c1} D_{c1} + U_{c1} N_{c1} = I_{n_{o1}}$ and $V_{c2} D_{c2} + U_{c2} N_{c2} = I_{n_{o2}}$, such that

$$\begin{bmatrix} \tilde{D}_{c1} & \tilde{N}_{c1} \\ -U_{c1} & V_{c1} \end{bmatrix} \begin{bmatrix} D_{11} & -N_{c1} \\ N_{11} & D_{c1} \end{bmatrix} = \begin{bmatrix} I_{n_{i1}} & 0 \\ 0 & I_{n_{o1}} \end{bmatrix} , \tag{4.3.22}$$

$$\begin{bmatrix} \tilde{D}_{c2} & \tilde{N}_{c2} \\ -U_{c2} & V_{c2} \end{bmatrix} \begin{bmatrix} D_{22} & -N_{c2} \\ N_{22} & D_{c2} \end{bmatrix} = \begin{bmatrix} I_{n_{i2}} & 0 \\ 0 & I_{n_{o2}} \end{bmatrix} . \tag{4.3.23}$$

Equations (4.3.22)-(4.3.23) are of the form $E_1 E_1^{-1} = I$, $E_2 E_2^{-1} = I$, where E_1 and E_2 are defined as:

$$E_1 := \begin{bmatrix} \tilde{D}_{c1} & \tilde{N}_{c1} \\ -U_{c1} & V_{c1} \end{bmatrix} , \quad E_2 := \begin{bmatrix} \tilde{D}_{c2} & \tilde{N}_{c2} \\ -U_{c2} & V_{c2} \end{bmatrix} . \tag{4.3.24}$$

The matrices in equation (4.3.22)-(4.3.23) are H–unimodular; hence the matrices $E_1 \in \mathrm{H}^{(n_{i1}+n_{o1}) \times (n_{i1}+n_{o1})}$ and $E_2 \in \mathrm{H}^{(n_{i2}+n_{o2}) \times (n_{i2}+n_{o2})}$ in equation (4.3.24) are H–unimodular. Now let

$$W_{12} := -U_{c1} D_{12} + V_{c1} N_{12} \quad , \quad W_{21} := -U_{c2} D_{21} + V_{c2} N_{21} \; . \tag{4.3.25}$$

With the H–unimodular matrices E_1 and E_2 defined as in equation (4.3.24) and $W_{12} \in \mathrm{H}^{n_{o1} \times n_{i2}}$ and $W_{21} \in \mathrm{H}^{n_{o2} \times n_{i1}}$ defined as in equation (4.3.25), by equations (4.3.21) and (4.3.22) we get:

$$\begin{bmatrix} E_1 & 0 \\ 0 & E_2 \end{bmatrix} \begin{bmatrix} D_{11} & D_{12} \\ N_{11} & N_{12} \\ D_{21} & D_{22} \\ N_{21} & N_{22} \end{bmatrix} = \begin{bmatrix} E_1 & 0 \\ 0 & E_2 \end{bmatrix} \begin{bmatrix} D_{p1} \\ N_{p1} \\ D_{p2} \\ N_{p2} \end{bmatrix} R = \begin{bmatrix} I_{n_{i1}} & 0 \\ 0 & W_{12} \\ 0 & I_{n_{i2}} \\ W_{21} & 0 \end{bmatrix} . \tag{4.3.26}$$

Equation (4.3.26) implies that any r.c.f.r. $(N_p \, , D_p)$ of P satisfies

$$\begin{bmatrix} D_{p1} \\ N_{p1} \\ D_{p2} \\ N_{p2} \end{bmatrix} = \begin{bmatrix} E_1^{-1} & 0 \\ 0 & E_2^{-1} \end{bmatrix} \begin{bmatrix} I_{n_{i1}} & 0 \\ 0 & W_{12} \\ 0 & I_{n_{i2}} \\ W_{21} & 0 \end{bmatrix} R^{-1} , \tag{4.3.27}$$

for some H–unimodular $E_1 \in \mathrm{H}^{(n_{i1}+n_{o1}) \times (n_{i1}+n_{o1})}$, $E_2 \in \mathrm{H}^{(n_{i2}+n_{o2}) \times (n_{i2}+n_{o2})}$, $R \in \mathrm{H}^{n_i \times n_i}$ and for some H–stable $W_{12} \in \mathrm{H}^{n_{o1} \times n_{i2}}$, $W_{21} \in \mathrm{H}^{n_{o2} \times n_{i1}}$. Therefore any r.c.f.r. $(N_p \, , D_p)$ of P satisfies conditions (4.3.3)-(4.3.4).

Statement (ii) is equivalent to statement (iii):

Suppose that statement (ii) holds. With E_1, E_2, R, W_{12} and W_{21} as in conditions (4.3.3)-(4.3.4), consider the following generalized Bezout identity, where $L \in \mathbf{H}^{n_o \times n_o}$ is any arbitrary $\mathbf{H}$–unimodular matrix:

$$\begin{bmatrix} R\begin{bmatrix} I_{n_{i1}} & 0 \\ 0 & 0 \end{bmatrix} E_1 & R\begin{bmatrix} 0 & 0 \\ I_{n_{i2}} & 0 \end{bmatrix} E_2 \\ L^{-1}\begin{bmatrix} 0 & I_{n_{o1}} \\ -W_{21} & 0 \end{bmatrix} E_1 & L^{-1}\begin{bmatrix} -W_{12} & 0 \\ 0 & I_{n_{o2}} \end{bmatrix} E_2 \end{bmatrix}$$

$$\cdot \begin{bmatrix} E_1^{-1}\begin{bmatrix} I_{n_{i1}} & 0 \\ 0 & W_{12} \end{bmatrix} R^{-1} & E_1^{-1}\begin{bmatrix} 0 & 0 \\ I_{n_{o1}} & 0 \end{bmatrix} L \\ E_2^{-1}\begin{bmatrix} 0 & I_{n_{i2}} \\ W_{21} & 0 \end{bmatrix} R^{-1} & E_2^{-1}\begin{bmatrix} 0 & 0 \\ 0 & I_{n_{o2}} \end{bmatrix} L \end{bmatrix} = \begin{bmatrix} \begin{bmatrix} I_{n_{i1}} & 0 \\ 0 & I_{n_{i2}} \end{bmatrix} & 0 \\ 0 & \begin{bmatrix} I_{n_{o1}} & 0 \\ 0 & I_{n_{o2}} \end{bmatrix} \end{bmatrix} . \tag{4.3.28}$$

In equation (4.3.28), let

$$\begin{bmatrix} -\tilde{N}_{p1} & \tilde{D}_{p1} \end{bmatrix} := L^{-1}\begin{bmatrix} 0 & I_{n_{o1}} \\ -W_{21} & 0 \end{bmatrix} E_1 ,$$

$$\begin{bmatrix} -\tilde{N}_{p2} & \tilde{D}_{p2} \end{bmatrix} := L^{-1}\begin{bmatrix} -W_{12} & 0 \\ 0 & I_{n_{o2}} \end{bmatrix} E_2 , \tag{4.3.29}$$

$$\tilde{N}_p := \begin{bmatrix} \tilde{N}_{p1} & \tilde{N}_{p2} \end{bmatrix} , \quad \tilde{D}_p := \begin{bmatrix} \tilde{D}_{p1} & \tilde{D}_{p2} \end{bmatrix} . \tag{4.3.30}$$

Now equation (4.3.28) is of the form

$$\begin{bmatrix} V_{p1} & U_{p1} & V_{p2} & U_{p2} \\ -\tilde{N}_{p1} & \tilde{D}_{p1} & -\tilde{N}_{p2} & \tilde{D}_{p2} \end{bmatrix} \begin{bmatrix} D_{p1} & -\tilde{U}_{p1} \\ N_{p1} & \tilde{V}_{p1} \\ D_{p2} & -\tilde{U}_{p2} \\ N_{p2} & \tilde{V}_{p2} \end{bmatrix} = \begin{bmatrix} I_{n_i} & 0 \\ 0 & I_{n_o} \end{bmatrix}, \quad (4.3.31)$$

where the H–stable matrices $V_{p1}, U_{p1}, V_{p2}, U_{p2}, \tilde{U}_{p1}, \tilde{V}_{p1}, \tilde{U}_{p2}, \tilde{V}_{p2}$ are defined in an obvious manner by comparing equations (4.3.28) and (4.3.31).

Now we must show that the pair $(\tilde{D}_p, \tilde{N}_p)$, defined by equations (4.3.29)-(4.3.30), is an l.c.f.r. of P : Since (4.3.28) is a generalized Bezout identity for the doubly-coprime pair $((N_p, D_p), (\tilde{D}_p, \tilde{N}_p))$, the pair $(\tilde{D}_p, \tilde{N}_p)$ is clearly l.c. Now since (N_p, D_p) is an r.c.f.r. of P, by definition, $\det D_p \in \mathrm{I}$; but by Lemma 2.4.4, since $((N_p, D_p), (\tilde{D}_p, \tilde{N}_p))$ is a doubly-coprime pair,

$$\det \tilde{D}_p \sim \det D_p \quad (4.3.32)$$

and hence, $\det \tilde{D}_p \in \mathrm{I}$. Now by equation (4.3.31), $-\tilde{N}_p D_p + \tilde{D}_p N_p = -\tilde{N}_{p1} D_{p1} + \tilde{D}_{p1} N_{p1} - \tilde{N}_{p2} D_{p2} + \tilde{D}_{p2} N_{p2} = 0$; therefore, $N_p D_p^{-1} = \tilde{D}_p^{-1} \tilde{N}_p$ and hence, $(\tilde{N}_p, \tilde{D}_p)$ is an l.c.f.r. of P. Since $L \in \mathrm{H}^{n_o \times n_o}$ is any arbitrary unimodular matrix, by Lemma 2.3.4 (ii), any l.c.f.r. of P is given by equation (4.3.29) and hence, conditions (4.3.5) and (4.3.6) are satisfied for any l.c.f.r. of P.

It is entirely similar to show that statement (iii) implies statement (ii); the proof once again follows from the generalized Bezout identity (4.3.28). □

Theorem 4.3.3 states that $P \in \mathrm{m}(\mathrm{G}_s)$ can be H–stabilized by a decentralized compensator C_d if and only if conditions (4.3.3)-(4.3.4) (equivalently, conditions (4.3.5)-(4.3.6)) are satisfied. So in Theorem 4.3.5 below, in order to find the class of all decentralized H–stabilizing compensators, we assume that any r.c.f.r. (N_p, D_p) and any l.c.f.r. $(\tilde{D}_p, \tilde{N}_p)$ of $P \in \mathrm{m}(\mathrm{G}_s)$ satisfy these conditions in addition to Assumption 4.2.1 (i).

Theorem 4.3.5. (Class of all decentralized H–stabilizing compensators in S(P , C_d))
Let P satisfy Assumption 4.2.1 (i); furthermore let $P \in \mathfrak{m}(G_s)$; let any r.c.f.r. (N_p , D_p) of P satisfy conditions (4.3.3) and (4.3.4) and equivalently, let any l.c.f.r. ($\tilde{D}_p$, $\tilde{N}_p$) of P satisfy conditions (4.3.5) and (4.3.6) of Theorem 4.3.3. Under these assumptions, the set $S_d(P)$ of *all* decentralized H–stabilizing compensators for P is given by

$$S_d(P) = \{ \; C_d = \begin{bmatrix} C_1 & 0 \\ 0 & C_2 \end{bmatrix} = \begin{bmatrix} \tilde{D}_{c1}^{-1}\tilde{N}_{c1} & 0 \\ 0 & \tilde{D}_{c2}^{-1}\tilde{N}_{c2} \end{bmatrix} \; |$$

$$\begin{bmatrix} \begin{bmatrix} \tilde{D}_{c1} & \tilde{N}_{c1} \end{bmatrix} & 0 \\ 0 & \begin{bmatrix} \tilde{D}_{c2} & \tilde{N}_{c2} \end{bmatrix} \end{bmatrix} = \begin{bmatrix} \begin{bmatrix} Q_{11} & Q_1 \end{bmatrix} E_1 & 0 \\ 0 & \begin{bmatrix} Q_{22} & Q_2 \end{bmatrix} E_2 \end{bmatrix},$$

$$Q_{11} \in H^{n_{i1} \times n_{i1}} \, , \, Q_1 \in H^{n_{i1} \times n_{o1}} \, , \, Q_2 \in H^{n_{i2} \times n_{o2}} \, , \, Q_{22} \in H^{n_{i2} \times n_{i2}}$$

$$\text{such that} \begin{bmatrix} Q_{11} & Q_1 W_{12} \\ Q_2 W_{21} & Q_{22} \end{bmatrix} \text{is H–unimodular } \} \, , \qquad (4.3.33)$$

equivalently,

$$S_d(P) = \{ \; C_d = \begin{bmatrix} C_1 & 0 \\ 0 & C_2 \end{bmatrix} = \begin{bmatrix} N_{c1}D_{c1}^{-1} & 0 \\ 0 & N_{c2}D_{c2}^{-1} \end{bmatrix} \; |$$

$$\begin{bmatrix} -N_{c1} & 0 \\ D_{c1} & 0 \\ \cdots & \cdots \\ 0 & -N_{c2} \\ 0 & D_{c2} \end{bmatrix} = \begin{bmatrix} E_1^{-1} \begin{bmatrix} -\hat{Q}_1 \\ \hat{Q}_{11} \end{bmatrix} & 0 \\ \cdots & \cdots \\ 0 & E_2^{-1} \begin{bmatrix} -\hat{Q}_2 \\ \hat{Q}_{22} \end{bmatrix} \end{bmatrix},$$

$$\hat{Q}_{11} \in H^{n_{o1} \times n_{o1}} \, , \, \hat{Q}_1 \in H^{n_{i1} \times n_{o1}} \, , \, \hat{Q}_2 \in H^{n_{i2} \times n_{o2}} \, , \, \hat{Q}_{22} \in H^{n_{o2} \times n_{o2}}$$

$$\text{such that} \begin{bmatrix} \hat{Q}_{11} & W_{12}\hat{Q}_2 \\ W_{21}\hat{Q}_1 & \hat{Q}_{22} \end{bmatrix} \text{is H–unimodular } \} \, . \qquad (4.3.34)$$

Comment 4.3.6

(i) Suppose that the plant P satisfies Assumptions 4.2.1 (i) and is block-triangular as in Comment 4.3.4 (v). Suppose that conditions (4.3.3)-(4.3.4) (equivalently, (4.3.5)-(4.3.6)) of Theorem 4.3.3 hold as explained in Comment 4.3.4 (v), where either the matrix W_{12} is zero (P is lower block-triangular) or the matrix W_{21} is zero (P is upper block-triangular); then the matrix

$$T := \begin{bmatrix} Q_{11} & Q_1 W_{12} \\ Q_2 W_{21} & Q_{22} \end{bmatrix} = \begin{bmatrix} Q_{11} & 0 \\ 0 & Q_{22} \end{bmatrix} + \begin{bmatrix} 0 & Q_1 \\ Q_2 & 0 \end{bmatrix} \begin{bmatrix} 0 & W_{21} \\ W_{12} & 0 \end{bmatrix}$$

$$\text{is } \mathbf{H}\text{–}unimodular \tag{4.3.35}$$

for all $Q_1, Q_2 \in \mathrm{m}(\mathbf{H})$ and for all $\mathbf{H}$–unimodular matrices $Q_{11} \in \mathbf{H}^{n_{i1} \times n_{i1}}$ and $Q_{22} \in \mathbf{H}^{n_{i2} \times n_{i2}}$. In this case, $(\tilde{D}_{c1}, \tilde{N}_{c1})$ and $(\tilde{D}_{c2}, \tilde{N}_{c2})$ are given by

$$\begin{bmatrix} \tilde{D}_{c1} & \tilde{N}_{c1} \end{bmatrix} = \begin{bmatrix} I_{n_{i1}} & Q_1 \end{bmatrix} E_1 \,, \quad \begin{bmatrix} \tilde{D}_{c2} & \tilde{N}_{c2} \end{bmatrix} = \begin{bmatrix} I_{n_{i2}} & Q_2 \end{bmatrix} E_2 \,. \tag{4.3.36}$$

Hence, if $P \in \mathrm{m}(\mathbf{G}_s)$ is block-triangular as in Comment 4.3.4 (v) and if conditions (4.3.3)-(4.3.4) of Theorem 4.3.3 hold, then the set $S_d(P)$ of all decentralized $\mathbf{H}$–stabilizing compensators in the expression (4.3.33) is *parametrized* by two *free* parameter-matrices Q_1 and $Q_2 \in \mathrm{m}(\mathbf{H})$. Similar comments hold for the expression in (4.3.34); if $P \in \mathrm{m}(\mathbf{G}_s)$ is block-triangular, then (N_{c1}, D_{c1}) and (N_{c2}, D_{c2}) are given by

$$\begin{bmatrix} -N_{c1} \\ D_{c1} \end{bmatrix} = E_1^{-1} \begin{bmatrix} -\hat{Q}_1 \\ I_{n_{o1}} \end{bmatrix} , \quad \begin{bmatrix} -N_{c2} \\ D_{c2} \end{bmatrix} = E_2^{-1} \begin{bmatrix} -\hat{Q}_2 \\ I_{n_{o2}} \end{bmatrix} , \tag{4.3.37}$$

where $\hat{Q}_1 \in \mathrm{m}(\mathbf{H})$, $\hat{Q}_2 \in \mathrm{m}(\mathbf{H})$ are any $\mathbf{H}$–stable matrices of appropriate dimensions.

(ii) The easiest way to choose $Q_{11}, Q_1, Q_2, Q_{22} \in \mathrm{m}(\mathbf{R_u})$ such that the matrix T in equation (4.3.35) is $\mathbf{H}$–unimodular is to choose either one (or both) of Q_1 and Q_2 as the zero matrix and then to choose both of Q_{11} and Q_{22} as arbitrary $\mathbf{H}$–unimodular matrices (or without loss of generality, as the identity matrices of size n_{i1} and n_{i2}, respectively).

(iii) In Theorem 4.3.5, if $P \in \mathbf{m}(\mathrm{G})$ instead of $\mathbf{m}(\mathrm{G}_s)$, then the matrices Q_{11}, Q_1, Q_2 and $Q_{22} \in \mathbf{m}(\mathrm{H})$ in the expression (4.3.33) should satisfy condition (4.3.35) *and* should be chosen so that

$$\det\tilde{D}_{c1} := \det(\begin{bmatrix} Q_{11} & Q_1 \end{bmatrix} E_1 \begin{bmatrix} I_{ni1} \\ 0 \end{bmatrix}) \in \mathrm{I}$$

$$\text{and} \quad \det\tilde{D}_{c2} := \det(\begin{bmatrix} Q_{22} & Q_2 \end{bmatrix} E_2 \begin{bmatrix} I_{ni2} \\ 0 \end{bmatrix}) \in \mathrm{I} . \tag{4.3.38}$$

(iv) Suppose that the plant $P \in \mathbf{m}(\mathrm{G}_s)$ is H–stable as in Comment 4.3.4 (iv); then by Theorem 4.3.5, $C_d = \begin{bmatrix} C_1 & 0 \\ 0 & C_2 \end{bmatrix}$ is a decentralized H–stablizing compensator for $P = \begin{bmatrix} P_{11} & P_{12} \\ P_{21} & P_{22} \end{bmatrix}$ if and only if

$$C_d = \begin{bmatrix} (Q_{11} - Q_1 P_{11})^{-1} Q_1 & 0 \\ 0 & (Q_{22} - Q_2 P_{22})^{-1} Q_2 \end{bmatrix}, \tag{4.3.39}$$

for some $Q_{11}, Q_{22}, Q_1, Q_2 \in \mathbf{m}(\mathrm{H})$ such that

$$\begin{bmatrix} Q_{11} & Q_1 P_{12} \\ Q_2 P_{21} & Q_{22} \end{bmatrix} \text{ is H–unimodular} . \tag{4.3.40}$$

Proof of Theorem 4.3.5

We only prove equation (4.3.33); the proof of (4.3.34) is similar.

We first show that, if C_d is given by the expression in equation (4.3.33), then C_d H–stabilizes P : With (N_p, D_p) as in conditions (4.3.3)-(4.3.4) and $(\tilde{D}_c, \tilde{N}_c)$ given as in the expression (4.3.33), we obtain

$$D_{H1} = \begin{bmatrix} \begin{bmatrix} \tilde{D}_{c1} & \tilde{N}_{c1} \end{bmatrix} & \vdots & 0 \\ 0 & \vdots & \begin{bmatrix} \tilde{D}_{c2} & \tilde{N}_{c2} \end{bmatrix} \end{bmatrix} \begin{bmatrix} D_{p1} \\ N_{p1} \\ \cdots \\ D_{p2} \\ N_{p2} \end{bmatrix}$$

$$= \begin{bmatrix} \begin{bmatrix} Q_{11} & Q_1 \end{bmatrix} E_1 & \vdots & 0 \\ 0 & \vdots & \begin{bmatrix} Q_{22} & Q_2 \end{bmatrix} E_2 \end{bmatrix} \begin{bmatrix} E_1^{-1} \begin{bmatrix} I_{n_{i1}} & 0 \\ 0 & W_{12} \end{bmatrix} \\ \cdots \\ E_2^{-1} \begin{bmatrix} 0 & I_{n_{i2}} \\ W_{21} & 0 \end{bmatrix} \end{bmatrix} R^{-1}$$

$$= \begin{bmatrix} Q_{11} & Q_2 W_{12} \\ Q_2 W_{21} & Q_{22} \end{bmatrix} R^{-1} = T R^{-1} . \qquad (4.3.41)$$

The matrix on the right-hand side of equation (4.3.41) is H–unimodular since by assumption, the matrix T in equation (4.3.35) and the matrix R in conditions (4.3.3)-(4.3.4) are H–unimodular. Therefore,

$$\tilde{D}_c D_p T^{-1} R + \tilde{N}_c N_p T^{-1} R = I_{n_i} . \qquad (4.3.42)$$

From equation (4.3.42),

$$\det\tilde{D}_c \det D_p = \det\left[I_{n_i} - \tilde{N}_c N_p T^{-1} R \right] \det R^{-1} \det T . \qquad (4.3.43)$$

Now $P \in \mathrm{m}(\mathrm{G}_s)$ implies that $N_p \in \mathrm{m}(\mathrm{G}_s)$ and therefore, $\tilde{N}_c N_p T^{-1} R \in \mathrm{m}(\mathrm{G}_s)$; hence, $\det\left[I_{n_i} - \tilde{N}_c N_p T^{-1} R \right] \in \mathrm{I}$. By equation (4.3.43), since $\det R^{-1} \det T \in \mathrm{J}$, we conclude that $\det\tilde{D}_c \det D_p \in \mathrm{I}$; hence, by Lemma 2.3.3 (ii), $\det D_p \in \mathrm{I}$ and $\det\tilde{D}_c \in \mathrm{I}$; but $\det\tilde{D}_c = \det\begin{bmatrix} \tilde{D}_{c1} & 0 \\ 0 & \tilde{D}_{c2} \end{bmatrix} \in \mathrm{I}$ if and only if $\det\tilde{D}_{c1} \in \mathrm{I}$ and $\det\tilde{D}_{c2} \in \mathrm{I}$, where

$$\tilde{D}_{c1} = \begin{bmatrix} Q_{11} & Q_1 \end{bmatrix} E_1 \begin{bmatrix} I_{n_{i1}} \\ 0 \end{bmatrix} ,$$

$$\tilde{D}_{c2} = \begin{bmatrix} Q_{22} & Q_2 \end{bmatrix} E_2 \begin{bmatrix} I_{n_{i2}} \\ 0 \end{bmatrix} . \qquad (4.3.44)$$

This proves that the l.c.f.r. given by the expression (4.3.33), namely

$$(\tilde{D}_{c1}, \tilde{N}_{c1}) = \begin{bmatrix} \tilde{D}_{c1} & \tilde{N}_{c1} \end{bmatrix} = \begin{bmatrix} Q_{11} & Q_1 \end{bmatrix} E_1 , \tag{4.3.45}$$

is an l.c.f.r. of C_1 and

$$(\tilde{D}_{c2}, \tilde{N}_{c2}) = \begin{bmatrix} \tilde{D}_{c2} & \tilde{N}_{c2} \end{bmatrix} = \begin{bmatrix} Q_{22} & Q_2 \end{bmatrix} E_2 , \tag{4.3.46}$$

is an l.c.f.r. of C_2. Now since equation (4.3.41) implies that D_{H1} is H–unimodular, the system $\mathbf{S}(P, C_d)$ is H–stable with this choice of decentralized compensator

$$C_d = \begin{bmatrix} C_1 & 0 \\ 0 & C_2 \end{bmatrix} = \begin{bmatrix} \tilde{D}_{c1}^{-1}\tilde{N}_{c1} & 0 \\ 0 & \tilde{D}_{c2}^{-1}\tilde{N}_{c2} \end{bmatrix} .$$

Now we show that any decentralized compensator C_d that H–stabilizes P is given by the expression in equation (4.3.33) for some $(Q_{11}, Q_1, Q_2, Q_{22}) \in \mathbf{m}(\mathrm{H})$ such that the matrix T in equation (4.3.35) is H–unimodular:

By assumption, $C_d = \begin{bmatrix} \tilde{D}_{c1}^{-1}\tilde{N}_{c1} & 0 \\ 0 & \tilde{D}_{c2}^{-1}\tilde{N}_{c2} \end{bmatrix}$ H–stabilizes P , where $\tilde{D}_{c1}, \tilde{D}_{c2}, \tilde{N}_{c1}$, $\tilde{N}_{c2} \in \mathbf{m}(\mathrm{H})$; hence by Theorem 4.2.5, the matrix D_{H1} in equation (4.2.19) is H–unimodular; equivalently,

$$\tilde{D}_c D_p D_{H1}^{-1} + \tilde{N}_c N_p D_{H1}^{-1} = I_{n_i} . \tag{4.3.47}$$

By Lemma 2.3.4 (i), since D_{H1} is H–unimodular, $(N_p D_{H1}^{-1} , D_p D_{H1}^{-1})$ is also an r.c.f.r. of P ; since conditions (4.3.3)-(4.3.4) are satisfied by assumption, equation (4.3.47) implies that there exists H–unimodular matrices E_1, E_2 and R such that the r.c.f.r. $(N_p D_{H1}^{-1} , D_p D_{H1}^{-1})$ of P satisfies

$$\begin{bmatrix} \tilde{D}_{c1} & \tilde{N}_{c1} & 0 & 0 \\ 0 & 0 & \tilde{D}_{c2} & \tilde{N}_{c2} \end{bmatrix} \begin{bmatrix} D_{p1} \\ N_{p1} \\ D_{p2} \\ N_{p2} \end{bmatrix} D_{H1}^{-1}$$

$$= \begin{bmatrix} \tilde{D}_{c1} & \tilde{N}_{c1} & 0 & 0 \\ 0 & 0 & \tilde{D}_{c2} & \tilde{N}_{c2} \end{bmatrix} \begin{bmatrix} E_1^{-1} \begin{bmatrix} I_{n_{i1}} & 0 \\ 0 & W_{12} \end{bmatrix} \\ \cdots \\ E_2^{-1} \begin{bmatrix} 0 & I_{n_{i2}} \\ W_{21} & 0 \end{bmatrix} \end{bmatrix} R^{-1} = I_{n_i} . \tag{4.3.48}$$

Define

$$Q_1 := \begin{bmatrix} \tilde{D}_{c1} & \tilde{N}_{c1} \end{bmatrix} E_1^{-1} \begin{bmatrix} 0 \\ I_{n_{o1}} \end{bmatrix} \quad \text{and} \quad Q_2 := \begin{bmatrix} \tilde{D}_{c2} & \tilde{N}_{c2} \end{bmatrix} E_2^{-1} \begin{bmatrix} 0 \\ I_{n_{o2}} \end{bmatrix} ; \tag{4.3.49}$$

clearly, Q_1 and $Q_2 \in \mathrm{m}(\mathrm{H})$; then by equations (4.3.48)-(4.3.49), we get

$$\begin{bmatrix} \begin{bmatrix} \tilde{D}_{c1} & \tilde{N}_{c1} \end{bmatrix} & 0 \\ 0 & \begin{bmatrix} \tilde{D}_{c2} & \tilde{N}_{c2} \end{bmatrix} \end{bmatrix} \begin{bmatrix} E_1^{-1} \begin{bmatrix} I_{n_{i1}} & 0 \\ 0 & W_{12} \end{bmatrix} R^{-1} & E_1^{-1} \begin{bmatrix} 0 & 0 \\ I_{n_{o1}} & 0 \end{bmatrix} L \\ E_2^{-1} \begin{bmatrix} 0 & I_{n_{i2}} \\ W_{21} & 0 \end{bmatrix} R^{-1} & E_2^{-1} \begin{bmatrix} 0 & 0 \\ 0 & I_{n_{o2}} \end{bmatrix} L \end{bmatrix}$$

$$= \begin{bmatrix} \begin{pmatrix} I_{n_{i1}} & 0 \\ 0 & I_{n_{i2}} \end{pmatrix} & \begin{pmatrix} Q_1 & 0 \\ 0 & Q_2 \end{pmatrix} L \end{bmatrix} . \tag{4.3.50}$$

Let the H–unimodular matrix R be partitioned as

$$R =: \begin{bmatrix} Q_{11} & R_{12} \\ R_{21} & Q_{22} \end{bmatrix} , \; Q_{11} \in \mathrm{H}^{n_{i1} \times n_{i1}} , \; R_{12} \in \mathrm{H}^{n_{i1} \times n_{i2}} ,$$

$$R_{21} \in \mathrm{H}^{n_{i2} \times n_{i1}} , \; Q_{22} \in \mathrm{H}^{n_{i2} \times n_{i2}} . \tag{4.3.51}$$

Post-multiply both sides of equation (4.3.50) by the first H–unimodular matrix in (4.3.28);

then

$$\begin{bmatrix} \begin{bmatrix} \tilde{D}_{c1} & \tilde{N}_{c1} \end{bmatrix} & 0 \\ 0 & \begin{bmatrix} \tilde{D}_{c2} & \tilde{N}_{c2} \end{bmatrix} \end{bmatrix} =$$

$$\begin{bmatrix} \begin{bmatrix} I_{n_{i1}} & 0 \\ 0 & I_{n_{i2}} \end{bmatrix} \begin{bmatrix} Q_1 & 0 \\ 0 & Q_2 \end{bmatrix} L \end{bmatrix} \begin{bmatrix} R \begin{bmatrix} I_{n_{i1}} & 0 \\ 0 & 0 \end{bmatrix} E_1 & R \begin{bmatrix} 0 & 0 \\ I_{n_{i2}} & 0 \end{bmatrix} E_2 \\ L^{-1} \begin{bmatrix} 0 & I_{n_{o1}} \\ -W_{21} & 0 \end{bmatrix} E_1 & L^{-1} \begin{bmatrix} -W_{12} & 0 \\ 0 & I_{n_{o2}} \end{bmatrix} E_2 \end{bmatrix}$$

$$= \begin{bmatrix} \begin{bmatrix} Q_{11} & Q_1 \end{bmatrix} E_1 & \begin{bmatrix} R_{12} - Q_1 W_{12} & 0 \end{bmatrix} E_2 \\ \begin{bmatrix} R_{21} - Q_2 W_{21} & 0 \end{bmatrix} E_1 & \begin{bmatrix} Q_{22} & Q_2 \end{bmatrix} E_2 \end{bmatrix} . \qquad (4.3.52)$$

Now by equation (4.3.52), since E_1 and E_2 are H–unimodular, $\begin{bmatrix} R_{12} - Q_1 W_{12} & 0 \end{bmatrix} E_2 = 0$ and $\begin{bmatrix} R_{21} - Q_2 W_{21} & 0 \end{bmatrix} E_1 = 0$ imply that

$$R_{12} = Q_1 W_{12} \quad \text{and} \quad R_{21} = Q_2 W_{21} \; ; \qquad (4.3.53)$$

therefore, by (4.3.51) and (4.3.53),

$$\begin{bmatrix} Q_{11} & Q_1 W_{12} \\ Q_2 W_{21} & Q_{22} \end{bmatrix} \in \mathrm{m}(\mathrm{H}) \text{ is H–unimodular.} \qquad (4.3.54)$$

We have hence shown that, by (4.3.52), $(\tilde{D}_{c1}, \tilde{N}_{c1})$ and $(\tilde{D}_{c2}, \tilde{N}_{c2})$ are of the form given by the expression in (4.3.33) for some H–stable matrices $Q_{11} \in \mathrm{H}^{n_{i1} \times n_{i1}}$, $Q_1 \in \mathrm{H}^{n_{i1} \times n_{o1}}$, $Q_2 \in \mathrm{H}^{n_{i2} \times n_{o2}}$, $Q_{22} \in \mathrm{H}^{n_{i2} \times n_{i2}}$ such that condition (4.3.35) (equivalently, (4.3.54)) holds. □

4.4 APPLICATION TO SYSTEMS REPRESENTED BY PROPER RATIONAL TRANSFER FUNCTIONS

In this section we consider the case where the principal ideal domain H is the ring R_u as in Section 2.2.

Consider the system $S(P, K_d)$ in Figure 4.6; let $K_d := \begin{bmatrix} K_1 & 0 \\ 0 & K_2 \end{bmatrix}$, $K_1 \in \mathbb{R}^{n_{i1} \times n_{o1}}$, $K_2 \in \mathbb{R}^{n_{i2} \times n_{o2}}$. The system $S(P, K_d)$ in Figure 4.6 is the same as $S(P, C_d)$, where the *dynamic* decentralized compensator $C_d = \begin{bmatrix} C_1 & 0 \\ 0 & C_2 \end{bmatrix} \in \mathrm{m}(G)$ of Figure 4.1 is replaced by the *real constant* decentralized compensator $K_d = \begin{bmatrix} K_1 & 0 \\ 0 & K_2 \end{bmatrix}$.

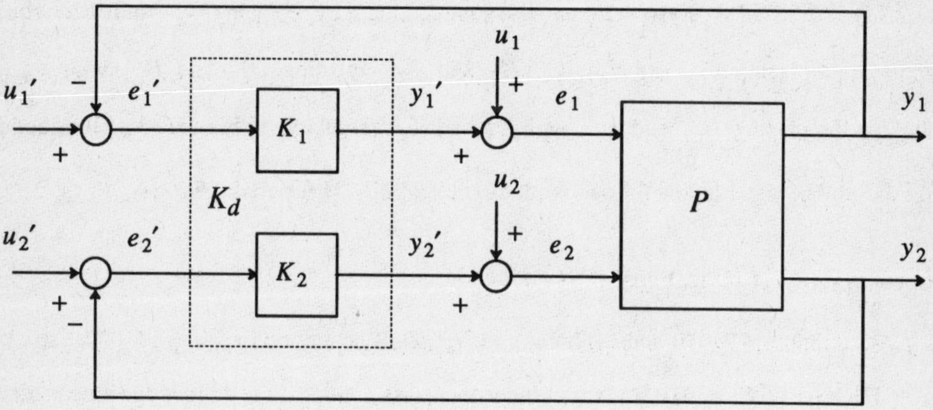

Figure 4.6. The constant-feedback decentralized control system $S(P, K_d)$.

Throughout this section we assume that Assumptions 4.2.1 of Section 4.2 hold, with H, G and G_s replaced by R_u, $\mathbb{R}_p(s)$ and $\mathbb{R}_{sp}(s)$, respectively. We also assume that the systems represented by the transfer matrices P, K and C_d have no hidden-modes associated with eigenvalues in $\bar{U}$.

Let (N_p, D_p) be any r.c.f.r. of P ; equations (4.2.10)-(4.2.11) describing the system $\mathbf{S}(P, C_d)$ as in Analysis 4.2.4 (i) are now replaced by equations (4.4.1)-(4.4.2) describing the system $\mathbf{S}(P, K_d)$ with constant decentralized feedback:

$$\begin{bmatrix} D_{p1} + K_1 N_{p1} \\ D_{p2} + K_2 N_{p2} \end{bmatrix} \xi_p = \begin{bmatrix} I_{ni1} & 0 & K_1 & 0 \\ 0 & I_{ni2} & 0 & K_2 \end{bmatrix} \begin{bmatrix} u_1 \\ u_2 \\ u_1' \\ u_2' \end{bmatrix} , \tag{4.4.1}$$

$$\begin{bmatrix} N_{p1} \\ N_{p2} \\ D_{p1} \\ D_{p2} \end{bmatrix} \xi_p = \begin{bmatrix} y_1 \\ y_2 \\ y_1' \\ y_2' \end{bmatrix} - \begin{bmatrix} 0 & 0 & 0 & 0 \\ 0 & 0 & 0 & 0 \\ -I_{ni1} & 0 & 0 & 0 \\ 0 & -I_{ni2} & 0 & 0 \end{bmatrix} \begin{bmatrix} u_1 \\ u_2 \\ u_1' \\ u_2' \end{bmatrix} . \tag{4.4.2}$$

Descriptions of $\mathbf{S}(P, K_d)$ analogous to equations (4.2.13)-(4.2.14), (4.2.15)-(4.2.16) and (4.2.17)-(4.2.18) of Analysis 4.2.4 are also easy to obtain by replacing $\tilde{D}_{c1}$ and $\tilde{D}_{c2}$ with I_{ni1} and I_{ni2}, replacing $\tilde{N}_{c1}$ and $\tilde{N}_{c2}$ with K_1 and K_2, replacing D_{c1} and D_{c2} with I_{no1} and I_{no2} and replacing N_{c1} and N_{c2} with K_1 and K_2, respectively, in each of these equations.

Corollary 4.4.1 below follows from the H–stability Theorem 4.2.5 for $\mathbf{S}(P, C_d)$:

Corollary 4.4.1. (H–stability of $\mathbf{S}(P, K_d)$)

Let Assumptions 4.2.1 (i) and (ii) hold; let (N_p, D_p) be any r.c.f.r., $(\tilde{D}_p, \tilde{N}_p)$ be any l.c.f.r. over $\mathfrak{m}(\mathrm{R_U})$ of $P \in \mathfrak{m}(\mathbb{R}_p(s))$. Under these assumptions, the following three statements are equivalent:

(i) $\mathbf{S}(P, K_d)$ is H–stable ;

(ii) $$\bar{D}_{H1} := \begin{bmatrix} D_p + K_d N_p \end{bmatrix} = \begin{bmatrix} D_{p1} + K_1 N_{p1} \\ D_{p2} + K_2 N_{p2} \end{bmatrix} \text{ is H–unimodular ;} \tag{4.4.3}$$

(iii) $\bar{D}_{H2} := \left[\tilde{D}_p + \tilde{N}_p K_d \right]$

$$= \left[\tilde{D}_{p1} + \tilde{N}_{p1} K_1 \quad \tilde{D}_{p2} + \tilde{N}_{p2} K_2 \right] \text{ is H–unimodular} . \qquad (4.4.4)$$

□

Note that each of statements (i) through (iii) of Corollary 4.4.1 implies that the system $\mathbf{S}(P, K_d)$ is well-posed.

Definition 4.4.2. (Decentralized fixed-eigenvalue)

The plant P is said to have a *decentralized fixed-eigenvalue* at $s_o \in \bar{\mathcal{U}}$ with respect to K_d iff $s_o \in \bar{\mathcal{U}}$ is a pole of the closed-loop I/O map $H_{\bar{y}\bar{u}}$ of the system $\mathbf{S}(P, K_d)$ *for all* $K_d \in \left\{ \begin{bmatrix} K_1 & 0 \\ 0 & K_2 \end{bmatrix} \mid K_1 \in \mathbb{R}^{n_{i1} \times n_{o1}}, K_2 \in \mathbb{R}^{n_{i2} \times n_{o2}} \right\}$.

Remark 4.4.3

Since we assume that the plant represented by the transfer matrix P has no hidden-modes associated with eigenvalues in $\bar{\mathcal{U}}$, it has no decentralized fixed-eigenvalues in $\bar{\mathcal{U}}$ corresponding to hidden-modes; it is clear that if P did have hidden-modes, these would remain as hidden-modes in the closed-loop system for all real constant-feedback and hence, be associated with decentralized fixed-eigenvalues.

Now $s_o \in \bar{\mathcal{U}}$ is a $\bar{\mathcal{U}}$–pole of the closed-loop I/O map $H_{\bar{y}\bar{u}}$ of $\mathbf{S}(P, K_d)$ if and only if $s_o \in \bar{\mathcal{U}}$ is a zero of a characteristic determinant of $\mathbf{S}(P, K_d)$. By Corollary 4.4.1, since $\bar{D}_{H1} \sim \bar{D}_{H2}$, given in equations (4.4.3)-(4.4.4), are characteristic determinants of the closed-loop I/O map $H_{\bar{y}\bar{u}}$ of the system $\mathbf{S}(P, K_d)$, $s_o \in \bar{\mathcal{U}}$ is a decentralized fixed-eigenvalue if and only if $\det\bar{D}_{H1}(s_o) = \det\bar{D}_{H2}(s_o) = 0$, for all $K_1, K_2 \in \mathfrak{m}(\mathbb{R})$; i.e.,

$$\det\begin{bmatrix} D_{p1}(s_o) + K_1 N_{p1}(s_o) \\ \\ D_{p2}(s_o) + K_2 N_{p2}(s_o) \end{bmatrix}$$

$$\sim \det\Big[\, \tilde{D}_{p1}(s_o) + \tilde{N}_{p1}(s_o)K_1 \quad \tilde{D}_{p2}(s_o) + \tilde{N}_{p2}(s_o)K_2 \,\Big] = 0$$

$$\text{for } \textit{all} \quad K_1 \in \mathbb{R}^{n_{i1} \times n_{o1}} \;, \quad K_2 \in \mathbb{R}^{n_{i2} \times n_{o2}} \;. \tag{4.4.5}$$

If $s_o \in \bar{\mathcal{U}}$ is a decentralized fixed-eigenvalue, then obviously $s_o \in \bar{\mathcal{U}}$ is an eigenvalue of the open-loop system P because with $K_1 = 0$, $K_2 = 0$, equation (4.4.5) becomes $\det\begin{bmatrix} D_{p1}(s_o) \\ D_{p2}(s_o) \end{bmatrix} = \det\Big[\, \tilde{D}_{p1}(s_o) \quad \tilde{D}_{p2}(s_o) \,\Big] = 0$ and hence, s_o is a zero of a characteristic determinant of P ; this eigenvalue at $s_o \in \bar{\mathcal{U}}$ remains a pole of the closed-loop I/O map $H_{\bar{y}\bar{u}}$ of the system $\mathbf{S}(P, K_d)$ for *all* real constant decentralized feedback compensators. We prefer to call such $s_o \in \bar{\mathcal{U}}$ a *fixed-eigenvalue* rather than a *fixed-mode*; although the eigenvalue at $s_o \in \bar{\mathcal{U}}$ remains fixed irrespective of the constant decentralized compensator, the *eigenvector* v_o associated with the fixed-eigenvalue $s_o \in \bar{\mathcal{U}}$ *depends* on K_1 and K_2. Therefore the "mode" $v_o e^{s_o t}$ may change "direction" depending on the choice of constant decentralized feedback; equivalently, the initial condition that sets up the mode $v_o e^{s_o t}$ varies with K_1, K_2 although the eigenvalue at $s_o \in \bar{\mathcal{U}}$ does not move. □

Theorem 4.3.3 gives the necessary and sufficient conditions on $P \in \mathfrak{m}(\mathbf{G}_s)$ for a decentralized $\mathbf{H}$–stabilizing compensator to exist. We now consider these necessary and sufficient conditions in the framework of $\mathbf{R}_{\mathcal{U}}$ instead of the general principal ideal domain $\mathbf{H}$; i.e., for the special case that the plant transfer function P is a strictly proper rational function.

Theorem 4.4.4. (Rank tests on $P = N_p D_p^{-1} = \tilde{D}_p^{-1} \tilde{N}_p$ for decentralized fixed-eigenvalues and $\mathbf{H}$–stabilizability)

Let P satisfy Assumption 4.2.1 (i); furthermore, let $P \in \mathfrak{m}(\mathbb{R}_{sp}(s))$; then the following six conditions are equivalent:

(i) There exists a decentralized H–stabilizing compensator C_d for P ;

(ii) Any r.c.f.r. (N_p , D_p) of P , partitioned as in equation (4.2.1), satisfies conditions (4.4.6) and (4.4.7) below:

$$E_1(s)\begin{bmatrix} D_{p1}(s) \\ N_{p1}(s) \end{bmatrix} R(s) = \begin{bmatrix} I_{n_{i1}} & 0 \\ 0 & W_{12}(s) \end{bmatrix}, \tag{4.4.6}$$

$$E_2(s)\begin{bmatrix} D_{p2}(s) \\ N_{p2}(s) \end{bmatrix} R(s) = \begin{bmatrix} 0 & I_{n_{i2}} \\ W_{21}(s) & 0 \end{bmatrix}, \tag{4.4.7}$$

(iii) Any l.c.f.r. $(\tilde{D}_p , \tilde{N}_p)$ of P , partitioned as in equation (4.2.2), satisfies conditions (4.4.8) and (4.4.9) below:

$$L(s)\begin{bmatrix} -\tilde{N}_{p1}(s) & \tilde{D}_{p1}(s) \end{bmatrix} E_1(s)^{-1} = \begin{bmatrix} 0 & I_{n_{o1}} \\ -W_{21}(s) & 0 \end{bmatrix}, \tag{4.4.8}$$

$$L(s)\begin{bmatrix} -\tilde{N}_{p2}(s) & \tilde{D}_{p2}(s) \end{bmatrix} E_2(s)^{-1} = \begin{bmatrix} -W_{12}(s) & 0 \\ 0 & I_{n_{o2}} \end{bmatrix}, \tag{4.4.9}$$

where $E_1(s) \in \mathrm{R_U}^{(n_{i1}+n_{o1})\times(n_{i1}+n_{o1})}$ is $\mathrm{R_U}$–unimodular, $E_2(s) \in \mathrm{R_U}^{(n_{i2}+n_{o2})\times(n_{i2}+n_{o2})}$ is $\mathrm{R_U}$–unimodular, $R \in \mathrm{R_U}^{n_i\times n_i}$ is $\mathrm{R_U}$–unimodular and $L \in \mathrm{R_U}^{n_o\times n_o}$ is $\mathrm{R_U}$–unimodular; the matrices $W_{12} \in \mathrm{R_U}^{n_{o1}\times n_{i2}}$ and $W_{21} \in \mathrm{R_U}^{n_{o2}\times n_{i1}}$ are $\mathrm{R_U}$–stable in equations (4.4.6) through (4.4.9).

(iv) Any r.c.f.r. (N_p, D_p) of P, partitioned as in equation (4.2.1), satisfies the rank conditions (4.4.10) and (4.4.11) below:

$$rank \begin{bmatrix} D_{p1}(s) \\ N_{p1}(s) \end{bmatrix} \geq n_{i1}, \quad \text{for all } s \in \bar{\mathcal{U}}, \tag{4.4.10}$$

$$rank \begin{bmatrix} D_{p2}(s) \\ N_{p2}(s) \end{bmatrix} \geq n_{i2}, \quad \text{for all } s \in \bar{\mathcal{U}}. \tag{4.4.11}$$

(v) Any l.c.f.r. $(\tilde{D}_p, \tilde{N}_p)$ of P, partitioned as in equation (4.2.2), satisfies the rank conditions (4.4.12) and (4.4.13) below:

$$rank \begin{bmatrix} -\tilde{N}_{p1}(s) & \tilde{D}_{p1}(s) \end{bmatrix} \geq n_{o1}, \quad \text{for all } s \in \bar{\mathcal{U}}, \tag{4.4.12}$$

$$rank \begin{bmatrix} -\tilde{N}_{p2}(s) & \tilde{D}_{p2}(s) \end{bmatrix} \geq n_{o2}, \quad \text{for all } s \in \bar{\mathcal{U}}. \tag{4.4.13}$$

(vi) The plant P has no decentralized fixed-eigenvalues in $\bar{\mathcal{U}}$.

Proof

We proved the equivalence of statements **(i)**, **(ii)**, **(iii)** in Theorem 4.3.3 for the general principal ideal domain $\mathbf{H}$; in Theorem 4.4.4 above, these three equivalent conditions are simply restated for the special case of the ring $\mathbf{R}_{\mathcal{U}}$ of stable rational functions. Here we first prove the equivalence of statement **(ii)** to statement **(iv)**; the equivalence of statements **(iii)** and **(v)** can be established similarly and we omit that proof. We then prove the equivalence of statements **(ii)** and **(vi)**.

Statement (ii) is equivalent to statement (iv):

(ii) => (iv) For any r.c.f.r. (N_p, D_p) of P, since the matrices $E_1(s), E_2(s), R(s)$ are

$\mathbf{R_U}$–unimodular, $rank\left(E_1(s)\begin{bmatrix} D_{p1}(s) \\ N_{p1}(s)\end{bmatrix} R(s)\right) = rank\begin{bmatrix} D_{p1}(s) \\ N_{p1}(s)\end{bmatrix} =$

$rank\begin{bmatrix} I_{n_{i1}} & 0 \\ 0 & W_{12}(s)\end{bmatrix} \geq n_{i1}$ and hence, condition (4.4.6) implies condition (4.4.10);

by the same reasoning, condition (4.4.7) implies condition (4.4.11).

(iv) => (ii) Condition (4.4.10) implies that there is an $(n_{i1}+n_{o1})\times(n_{i1}+n_{o1})$ $\mathbf{R_U}$–unimodular matrix L_1 and an $n_i \times n_i$ $\mathbf{R_U}$–unimodular matrix R_1 such that

$$L_1\begin{bmatrix} D_{p1} \\ N_{p1}\end{bmatrix} R_1 = \begin{bmatrix} I_{n_{i1}} & 0 \\ 0 & \hat{N}_{12}\end{bmatrix}, \tag{4.4.15}$$

where $\hat{N}_{12} \in \mathbf{R_U}^{n_{o1}\times n_{i2}}$ is some $\mathbf{R_U}$–stable matrix.

Condition (4.4.11) on the other hand, implies that there is an $(n_{i2}+n_{o2})\times(n_{i2}+n_{o2})$ $\mathbf{R_U}$–unimodular matrix L_2 (corresponding to elementary row operations in $\mathbf{R_U}$) such that

$$L_2\left(\begin{bmatrix} D_{p2} \\ N_{p2}\end{bmatrix} R_1\right) = \begin{bmatrix} -\hat{D}_{21} & \hat{D}_{22} \\ \hat{N}_{21} & 0\end{bmatrix}, \tag{4.4.16}$$

where $\hat{D}_{22} \in \mathbf{R_U}^{n_{i2}\times n_{i2}}$, $\hat{D}_{21} \in \mathbf{R_U}^{n_{i2}\times n_{i1}}$, $\hat{N}_{21} \in \mathbf{R_U}^{n_{o2}\times n_{i1}}$ are some $\mathbf{R_U}$–stable matrices and $\hat{D}_{21}$, $\hat{D}_{22}$ also satisfy

$$rank\begin{bmatrix} \hat{D}_{21} & \hat{D}_{22}\end{bmatrix} = n_{i2}, \quad \text{for all } s \in \bar{\mathcal{U}}. \tag{4.4.17}$$

By Lemma 2.6.1 (ii). equation (4.4.17) implies that the pair $(\hat{D}_{22}, \hat{D}_{21})$ is l.c.; hence, (recalling the generalized Bezout identities in Corollary 2.3.8) there exist matrices V_{2l}, U_{2l}, X_2, $Y_2, U_2, V_2 \in \mathbf{m}(\mathbf{R_U})$ such that

$$\begin{bmatrix} V_2 & U_2 \\ -\hat{D}_{21} & \hat{D}_{22} \end{bmatrix} \begin{bmatrix} Y_2 & -U_{21} \\ X_2 & V_{21} \end{bmatrix} = \begin{bmatrix} I_{n_{i1}} & 0 \\ 0 & I_{n_{i2}} \end{bmatrix} . \tag{4.4.18}$$

Now since (N_p, D_p) is an r.c. pair, by Lemma 2.6.1 (i), $rank \begin{bmatrix} D_p \\ N_p \end{bmatrix} = n_i$. But since L_1 and L_2 are $\mathbf{R_u}$–unimodular matrices,

$$rank \left(\begin{bmatrix} L_1 & \vdots & 0 \\ \cdots & & \cdots \\ 0 & \vdots & L_2 \end{bmatrix} \begin{bmatrix} D_{p1}(s) \\ N_{p1}(s) \\ \cdots \\ D_{p2}(s) \\ N_{p2}(s) \end{bmatrix} R_1 \right)$$

$$= rank \begin{bmatrix} I_{n_{i1}} & 0 \\ 0 & \hat{N}_{12}(s) \\ \cdots & \cdots \\ \hat{D}_{21}(s) & \hat{D}_{22}(s) \\ \hat{N}_{21}(s) & 0 \end{bmatrix} = n_i \quad , \text{ for all } s \in \bar{\mathcal{U}} \; ; \tag{4.4.19}$$

hence,

$$rank \begin{bmatrix} \hat{N}_{12}(s) \\ \hat{D}_{22}(s) \end{bmatrix} = n_{i2} \text{ , for all } s \in \bar{\mathcal{U}} \, . \tag{4.4.20}$$

By Lemma 2.6.1 (i), equation (4.4.20) implies that $(\hat{N}_{12}, \hat{D}_{22})$ is an r.c. pair, and hence, (recalling the generalized Bezout identities in Corollary 2.3.8) there exist matrices V_{2r} , U_{2r} , $\tilde{X}_2$, $\tilde{Y}_2$, $\tilde{U}_2$, $\tilde{V}_2 \in \mathfrak{m}(\mathbf{R_u})$ such that

$$\begin{bmatrix} V_{2r} & U_{2r} \\ -\tilde{X}_2 & \tilde{Y}_2 \end{bmatrix} \begin{bmatrix} \hat{D}_{22} & -\tilde{U}_2 \\ \hat{N}_{12} & \tilde{V}_2 \end{bmatrix} = \begin{bmatrix} I_{n_{i2}} & 0 \\ 0 & I_{n_{o1}} \end{bmatrix} . \tag{4.4.21}$$

Using the two generalized Bezout identities (4.4.18) and (4.4.21), it can be easily verified that equation (4.2.22) below holds:

$$\begin{bmatrix} V_2 + U_2 V_{2r} \hat{D}_{21} & U_2 U_{2r} \\ -\tilde{X}_2 \hat{D}_{21} & \tilde{Y}_2 \end{bmatrix} \begin{bmatrix} Y_2 & -U_{2l} \tilde{U}_2 \\ \hat{N}_{12} X_2 & \tilde{V}_2 + \hat{N}_{12} V_{2l} \tilde{U}_2 \end{bmatrix} = \begin{bmatrix} I_{n_{i1}} & 0 \\ 0 & I_{n_{o1}} \end{bmatrix} . \tag{4.4.22}$$

Now let

$$R_2 := \begin{bmatrix} Y_2 & -U_{2l} \\ X_2 & V_{2l} \end{bmatrix} \in \mathrm{R_u}^{n_i \times n_i} ; \tag{4.4.23}$$

then R_2 is $\mathrm{R_u}$–unimodular by equation (4.4.18). Let

$$R := R_1 R_2 \begin{bmatrix} I_{n_{i1}} & U_2 V_{2r} \\ 0 & I_{n_{i2}} \end{bmatrix} \in \mathrm{R_u}^{n_i \times n_i} ; \tag{4.4.24}$$

then R is $\mathrm{R_u}$–unimodular by (4.4.15) and (4.4.23). Now let

$$E_1 := \begin{bmatrix} V_2 + U_2 V_{2r} \hat{D}_{21} & U_2 U_{2r} \\ -\tilde{X}_2 \hat{D}_{21} & \tilde{Y}_2 \end{bmatrix} L_1 \in \mathrm{R_u}^{(n_{i1}+n_{o1}) \times (n_{i1}+n_{o1})} ; \tag{4.4.25}$$

then E_1 is $\mathrm{R_u}$–unimodular by equations (4.4.22) and (4.4.15). Let

$$E_2 := \begin{bmatrix} I_{n_{i2}} & 0 \\ \hat{N}_{21} U_{2l} \tilde{U}_2 \tilde{X}_2 & I_{n_{o2}} \end{bmatrix} L_2 \in \mathrm{R_u}^{(n_{i2}+n_{o2}) \times (n_{i2}+n_{o2})} ; \tag{4.4.26}$$

then E_2 is also $\mathrm{R_u}$–unimodular by equation (4.4.16). Now let

$$W_{12} := \tilde{X}_2 \in \mathrm{R_u}^{n_{o1} \times n_{i2}} \quad \text{and} \quad W_{21} := \hat{N}_{21} Y_2 \in \mathrm{R_u}^{n_{o2} \times n_{i1}} . \tag{4.4.27}$$

Then from equations (4.4.23)-(4.4.27) we obtain

$$\begin{bmatrix} E_1 & 0 \\ \cdots & \cdots \\ 0 & E_2 \end{bmatrix} \begin{bmatrix} D_{p1} \\ N_{p1} \\ \cdots \\ D_{p2} \\ N_{p2} \end{bmatrix} R = \begin{bmatrix} I_{n_{i1}} & 0 \\ 0 & W_{12} \\ \cdots & \cdots \\ 0 & I_{n_{i2}} \\ W_{21} & 0 \end{bmatrix} . \tag{4.4.28}$$

Equation (4.4.28) implies that for any r.c.f.r. (N_p, D_p) of P, conditions (4.4.6) and (4.4.7) are satisfied for some $\mathbb{R}_{\mathcal{U}}$–unimodular matrices E_1, E_2 and R.

Statement (ii) is equivalent to statement (vi):

(vi) => (ii) We will show that if statement (ii) does *not* hold, then statement (vi) does *not* hold either. We proved above that statement (ii) is equivalent to statement (iv); therefore if statement (ii) fails then statement (iv) also fails. Suppose, without loss of generality, that condition (4.4.10) fails; i.e., there is an $s_o \in \bar{\mathcal{U}}$ such that

$$rank \begin{bmatrix} D_{p1}(s_o) \\ N_{p1}(s_o) \end{bmatrix} < n_{i1} . \tag{4.4.29}$$

Equation (4.4.29) implies that for all $K_1 \in \mathbb{R}^{n_{i1} \times n_{o1}}$,

$$rank \begin{bmatrix} D_{p1}(s_o) + K_1 N_{p1}(s_o) \end{bmatrix} = rank \left(\begin{bmatrix} I_{n_{i1}} & K_1 \end{bmatrix} \begin{bmatrix} D_{p1}(s_o) \\ N_{p1}(s_o) \end{bmatrix} \right)$$

$$\leq rank \begin{bmatrix} D_{p1}(s_o) \\ N_{p1}(s_o) \end{bmatrix} < n_{i1} . \tag{4.4.30}$$

But equation (4.4.30) implies that for all $K_1 \in \mathbb{R}^{n_{i1} \times n_{o1}}$, $K_2 \in \mathbb{R}^{n_{i2} \times n_{o2}}$,

$$rank \begin{bmatrix} D_{p1}(s_o) + K_1 N_{p1}(s_o) \\ D_{p2}(s_o) + K_2 N_{p2}(s_o) \end{bmatrix}$$

$$\leq \; rank \left[D_{p1}(s_o) + K_1 N_{p1}(s_o) \right] \; + rank \left[D_{p2}(s_o) + K_2 N_{p2}(s_o) \right] \; < n_{i1} + n_{i2} \,, \tag{4.4.31}$$

and hence, by equation (4.4.5) of Remark 4.4.3, this $s_o \in \bar{\mathcal{U}}$ is a decentralized fixed-eigenvalue; therefore, statement (vi) fails.

(ii) => (vi) By assumption, conditions (4.4.6) and (4.4.7) hold for any r.c.f.r. (N_p, D_p) of P ; suppose now, for a contradiction, that P has a decentralized fixed eigenvalue at $s_o \in \bar{\mathcal{U}}$. Then by equation (4.4.5) of Remark 4.4.3, since the matrix $R \in \mathbf{R_U}^{n_i \times n_i}$ in conditions (4.4.6)-(4.4.7) is $\mathbf{R_U}$–unimodular, for all $K_1 \in \mathbb{R}^{n_{i1} \times n_{o1}}$, $K_2 \in \mathbb{R}^{n_{i2} \times n_{o2}}$,

$$rank \begin{bmatrix} D_{p1}(s_o) + K_1 N_{p1}(s_o) \\ D_{p2}(s_o) + K_2 N_{p2}(s_o) \end{bmatrix} R(s_o) = rank \begin{bmatrix} D_{p1}(s_o) + K_1 N_{p1}(s_o) \\ D_{p2}(s_o) + K_2 N_{p2}(s_o) \end{bmatrix} < n_i \,. \tag{4.4.32}$$

Now let

$$\begin{bmatrix} D_{p1}(s_o) \\ N_{p1}(s_o) \end{bmatrix} R(s_o) =: \begin{bmatrix} D_{11}(s_o) & D_{12}(s_o) \\ N_{11}(s_o) & N_{12}(s_o) \end{bmatrix} . \tag{4.4.33}$$

Since the matrix $E_1 \in \mathbf{R_U}^{(n_{i1}+n_{o1}) \times (n_{i1}+n_{o1})}$ in condition (4.4.6) is $\mathbf{R_U}$–unimodular, by Lemma 2.6.1 (i), the pair (N_{11}, D_{11}) is r.c. because

$$rank \begin{bmatrix} D_{11}(s_o) \\ N_{11}(s_o) \end{bmatrix} = rank \left(E_1(s_o) \begin{bmatrix} D_{11}(s_o) \\ N_{11}(s_o) \end{bmatrix} \right) = rank \begin{bmatrix} I_{n_{i1}} \\ 0 \end{bmatrix} = n_{i1} \,. \tag{4.4.34}$$

Now by Corollary 2.6.3 (ii), equation (4.4.34) implies that there exists a real constant matrix $\hat{K}_1 \in \mathbb{R}^{n_{i1} \times n_{o1}}$ such that

$$rank \left[D_{11}(s_o) + \hat{K}_1 N_{11}(s_o) \right] = n_{i1} \,; \tag{4.4.35}$$

i.e., the complex matrix $(D_{11}(s_o) + \hat{K}_1 N_{11}(s_o)) \in \mathbb{C}^{n_{i1} \times n_{i1}}$ is nonsingular. Let

$$L_1 := (\, D_{11}(s_o) + \hat{K}_1 N_{11}(s_o)\,)^{-1} \quad \text{and} \quad R_1 := \begin{bmatrix} I_{n_{i1}} & -L_1(D_{12} + \hat{K}_1 N_{12})(s_o) \\ 0 & I_{n_{i2}} \end{bmatrix}; \tag{4.4.36}$$

note that $L_1 \in \mathbb{C}^{n_{i1} \times n_{i1}}$ and $R_1 \in \mathbb{C}^{n_i \times n_i}$ are nonsingular. By equations (4.4.33) and (4.4.36),

$$L_1 \Big[\, D_{p1}(s_o) + \hat{K}_1 N_{p1}(s_o) \,\Big] R(s_o) R_1$$

$$= L_1 \Big[\, D_{11}(s_o) + \hat{K}_1 N_{11}(s_o) \quad D_{12}(s_o) + \hat{K}_1 N_{12}(s_o) \,\Big] R_1 = \Big[\, I_{n_{i1}} \quad 0 \,\Big]\,. \tag{4.4.37}$$

Now let

$$\begin{bmatrix} L_1(D_{p1}(s_o) + \hat{K}_1 N_{p1}(s_o)) \\ D_{p2}(s_o) \\ N_{p2}(s_o) \end{bmatrix} R(s_o) R_1 =: \begin{bmatrix} I_{n_{i1}} & 0 \\ D_{21}(s_o) & D_{22}(s_o) \\ N_{21}(s_o) & N_{22}(s_o) \end{bmatrix}. \tag{4.4.38}$$

Since L_1 and R_1 are nonsingular complex matrices, equations (4.4.32) and (4.4.38) imply that for all $K_2 \in \mathbb{R}^{n_{i2} \times n_{o2}}$,

$$rank \begin{bmatrix} D_{p1}(s_o) + \hat{K}_1 N_{p1}(s_o) \\ D_{p2}(s_o) + K_2 N_{p2}(s_o) \end{bmatrix} R(s_o)$$

$$= rank \begin{bmatrix} L_1(D_{p1}(s_o) + \hat{K}_1 N_{p1}(s_o)) R(s_o) R_1 \\ (D_{p2}(s_o) + K_2 N_{p2}(s_o)) R(s_o) R_1 \end{bmatrix}$$

$$= rank \left(\begin{bmatrix} I_{n_{i1}} & 0 & 0 \\ 0 & I_{n_{i2}} & K_2 \end{bmatrix} \begin{bmatrix} I_{n_{i1}} & 0 \\ D_{21}(s_o) & D_{22}(s_o) \\ N_{21}(s_o) & N_{22}(s_o) \end{bmatrix} \right)$$

$$= n_{i1} + rank \Big[\, D_{22}(s_o) + K_2 N_{22}(s_o) \,\Big] < n_i\,. \tag{4.4.39}$$

But equation (4.4.39) implies that for all $K_2 \in \mathbb{R}^{n_{i2} \times n_{o2}}$,

$$rank \left[D_{22}(s_o) + K_2 N_{22}(s_o) \right] < n_{i2} , \tag{4.4.40}$$

and hence, by Lemma 2.6.2 (i),

$$\max_{K_2 \in \mathfrak{m}(\mathbb{R})} rank \left[D_{22}(s_o) + K_2 N_{22}(s_o) \right] = rank \begin{bmatrix} D_{22}(s_o) \\ N_{22}(s_o) \end{bmatrix} < n_{i2} . \tag{4.4.41}$$

Equation (4.4.41) implies that

$$rank \begin{bmatrix} L_1(D_{p1}(s_o) + \hat{K}_1 N_{p1}(s_o)) \\ D_{p2}(s_o) \\ N_{p2}(s_o) \end{bmatrix} R(s_o) R_1 = rank \begin{bmatrix} I_{n_{i1}} & 0 \\ D_{21}(s_o) & D_{22}(s_o) \\ N_{21}(s_o) & N_{22}(s_o) \end{bmatrix}$$

$$< n_{i1} + n_{i2} . \tag{4.4.42}$$

But the matrix $R_1 \in \mathbb{C}^{n_i \times n_i}$ is nonsingular and the matrix $E_2 \in \mathrm{R_U}^{(n_{i2}+n_{o2}) \times (n_{i2}+n_{o2})}$ in condition (4.4.7) is $\mathrm{R_U}$–unimodular; therefore, condition (4.4.7) and equation (4.4.42) imply that

$$rank \left(\begin{bmatrix} I_{n_{i1}} & 0 \\ \cdots & \cdots \\ 0 & E_2 \end{bmatrix} \begin{bmatrix} L_1(D_{p1}(s_o) + \hat{K}_1 N_{p1}(s_o)) \\ \cdots \\ D_{p2}(s_o) \\ N_{p2}(s_o) \end{bmatrix} R(s_o) \right.$$

$$= rank \begin{bmatrix} I_{n_{i1}} & L_1(D_{12}(s_o) + \hat{K}_1 N_{12}(s_o)) \\ \cdots & \cdots \\ 0 & I_{n_{i2}} \\ W_{21}(s_o) & 0 \end{bmatrix} < n_{i1} + n_{i2} . \tag{4.4.43}$$

This is clearly a contradiction since by elementary row operations on the second matrix in equation (4.4.43), it is easy to see that the rank of the matrix in (4.4.43) should be exactly equal to $n_{i1} + n_{i2}$. We conclude that whenever conditions (4.4.6) and (4.4.7) hold, the plant P can *not* have any decentralized fixed-eigenvalues in $\bar{\mathrm{U}}$. □

Corollary 4.4.5. (Rank test on $P = N_{pr} D^{-1} N_{pl} + G$ for decentralized fixed-eigenvalues and H–stabilizability)

Let P satisfy Assumption 4.2.1 (i); furthermore, let $P \in \mathfrak{m}(\mathbb{R}_{sp}(s))$; then the following three conditions are equivalent:

(i) Any b.c.f.r. (N_{pr}, D, N_{pl}, G) of P, partitioned as in equation (4.2.3), satisfies the rank conditions (4.4.44) and (4.4.45) below:

$$rank \begin{bmatrix} D(s) & -N_{pl2}(s) \\ N_{pr1}(s) & G_{12} \end{bmatrix} \geq n \ , \quad \text{for all } s \in \bar{\mathcal{U}} \ , \qquad (4.4.44)$$

$$rank \begin{bmatrix} D(s) & -N_{pl1}(s) \\ N_{pr2}(s) & G_{21} \end{bmatrix} \geq n \ , \quad \text{for all } s \in \bar{\mathcal{U}} \ . \qquad (4.4.45)$$

(ii) The plant P has no decentralized fixed-eigenvalues in $\bar{\mathcal{U}}$.

(iii) There exists a decentralized H–stabilizing compensator C_d for P .

Proof

We prove that conditions (4.4.44) and (4.4.45) are equivalent to conditions (4.4.6) and (4.4.7) of Theorem 4.4.4; the equivalence of statements (i), (ii) and (iii) of Corollary 4.4.5 follow. Note that the equivalent conditions (i) and (ii) of Corollary 4.4.5 are equivalent to (4.4.6) and (4.4.7) for *all* plants $P \in \mathfrak{m}(\mathbb{R}_p(s))$; the assumption that P is strictly proper is only needed to establish that these conditions are in turn equivalent to condition (iii).

By Theorem 4.3.3 applied to the r.c.f.r. $(N_{pr} X + G Y, Y)$ of $P \in \mathfrak{m}(\mathbb{R}_{sp}(s))$ as in Comment 4.3.4 (vii), condition (4.4.6) is equivalent to

$$E_1(s) \begin{bmatrix} Y_1(s) \\ N_{pr1}(s) X(s) + G_{11}(s) Y_1(s) + G_{12}(s) Y_2(s) \end{bmatrix} R(s)$$

$$= \begin{bmatrix} I_{n_{i1}} & 0 \\ 0 & W_{12}(s) \end{bmatrix} \tag{4.4.46}$$

and condition (4.4.7) is equivalent to

$$E_2(s) \begin{bmatrix} Y_2(s) \\ N_{pr2}(s)X(s) + G_{21}(s)Y_1(s) + G_{22}(s)Y_2(s) \end{bmatrix} R(s)$$

$$= \begin{bmatrix} 0 & I_{n_{i2}} \\ W_{21}(s) & 0 \end{bmatrix}, \tag{4.4.47}$$

where $E_1 \in \mathbf{R_u}^{(n_{i1}+n_{o1})\times(n_{i1}+n_{o1})}$, $E_2 \in \mathbf{R_u}^{(n_{i2}+n_{o2})\times(n_{i2}+n_{o2})}$ and $R \in \mathbf{R_u}^{n_i \times n_i}$ are $\mathbf{R_u}$–unimodular; $W_{12} \in \mathbf{R_u}^{n_{o1}\times n_{i2}}$, $W_{21} \in \mathbf{R_u}^{n_{o2}\times n_{i1}}$. Now by Theorem 4.4.4, condition (4.4.6) is equivalent to condition (4.4.10); therefore, by (4.4.46), since E_1 and R are $\mathbf{R_u}$–unimodular matrices,

$$rank\, E_1(s) \begin{bmatrix} Y_1(s) \\ N_{pr1}(s)X(s) + G_{11}(s)Y_1(s) + G_{12}(s)Y_2(s) \end{bmatrix} R(s)$$

$$= rank \begin{bmatrix} Y_1(s) \\ N_{pr1}(s)X(s) + G_{11}(s)Y_1(s) + G_{12}(s)Y_2(s) \end{bmatrix} \geq n_{i1},$$

$$\text{for all } s \in \bar{\mathcal{U}}. \tag{4.4.48}$$

Now from the generalized Bezout identity (4.2.5),

$$\begin{bmatrix} D & -N_{pl1} & -N_{pl2} \\ 0 & I_{n_{i1}} & 0 \\ N_{pr1} & G_{11} & G_{12} \end{bmatrix} \begin{bmatrix} V_{pl} & X \\ -U_{pl1} & Y_1 \\ -U_{pl2} & Y_2 \end{bmatrix}$$

$$= \begin{bmatrix} I_n & 0 \\ -U_{pl1} & Y_1 \\ N_{pr1}V_{pl} - G_{11}U_{pl1} - G_{12}U_{pl2} & N_{pr1}X + G_{11}Y_1 + G_{12}Y_2 \end{bmatrix} . \quad (4.4.49)$$

Since the second matrix on the left-hand side of equation (4.4.49) is $\mathbf{R_U}$–unimodular by (4.2.5), for all $s \in \bar{\mathrm{U}}$, the rank of the first matrix on the left-hand side is the same as the rank of the matrix on the right-hand side; therefore, by equation (4.4.48),

$$rank \begin{bmatrix} D(s) & -N_{pl1}(s) & -N_{pl2}(s) \\ 0 & I_{n_{i1}} & 0 \\ N_{pr1}(s) & G_{11}(s) & G_{12}(s) \end{bmatrix}$$

$$= n + rank \begin{bmatrix} Y_1(s) \\ N_{pr1}(s)X(s) + G_{11}(s)Y_1(s) + G_{12}(s)Y_2(s) \end{bmatrix} \geq n + n_{i1} ,$$

$$\text{for all} \quad s \in \bar{\mathrm{U}} . \quad (4.4.50)$$

By elementary row operations on the first matrix in equation (4.4.50), it is easy to see that

$$rank \begin{bmatrix} D(s) & -N_{pl1}(s) & -N_{pl2}(s) \\ 0 & I_{n_{i1}} & 0 \\ N_{pr1}(s) & G_{11}(s) & G_{12}(s) \end{bmatrix}$$

$$= rank \begin{bmatrix} D(s) & -N_{pl2}(s) \\ N_{pr1}(s) & G_{12} \end{bmatrix} + n_{i1} \geq n + n_{i1} , \quad \text{for all} \quad s \in \bar{\mathrm{U}} . \quad (4.4.51)$$

By equation (4.4.51), we conclude that condition (4.4.6) is equivalent to condition (4.4.44). The equivalence of condition (4.4.7) to condition (4.4.45) can be established similarly using equation (4.4.47).

Remark 4.4.6. (Rank test on $P = \bar{C}(sI_n - \bar{A})^{-1}\bar{B} + \bar{E}$ for decentralized fixed-eigenvalues and H–stabilizability)

Let P satisfy Assumption 4.2.1 (i); let $(\bar{A}, \bar{B}, \bar{C}, \bar{E})$ be the state-space description of $P\,\mathrm{m}(\mathbb{R}_p(s))$ given in Example 2.4.3. Following Example 2.4.3, (N_{pr}, D, N_{pl}, G) is a b.c.f.r. of P, where

$$N_{pr} := \begin{bmatrix} (s+a)^{-1}\bar{C}_1 \\ (s+a)^{-1}\bar{C}_2 \end{bmatrix}, \quad D := \left[(s+a)^{-1}(sI_n - \bar{A}) \right],$$

$$N_{pl} := \begin{bmatrix} \bar{B}_1 & \bar{B}_2 \end{bmatrix}, \quad G := \begin{bmatrix} \bar{E}_{11} & \bar{E}_{12} \\ \bar{E}_{21} & \bar{E}_{22} \end{bmatrix}, \tag{4.4.52}$$

where $-a \in \mathbb{R} \cap \mathbb{C}\backslash \mathcal{U}$, $\bar{A} \in \mathbb{R}^{n \times n}$, $\bar{C}_1 \in \mathbb{R}^{n_{o1} \times n}$, $\bar{C}_2 \in \mathbb{R}^{n_{o2} \times n}$, $\bar{B}_1 \in \mathbb{R}^{n \times n_{i1}}$, $\bar{B}_2 \in \mathbb{R}^{n \times n_{i2}}$, $\bar{E}_{11} \in \mathbb{R}^{n_{o1} \times n_{i1}}$, $\bar{E}_{12} \in \mathbb{R}^{n_{o1} \times n_{i2}}$, $\bar{E}_{21} \in \mathbb{R}^{n_{o2} \times n_{i1}}$, $\bar{E}_{22} \in \mathbb{R}^{n_{o2} \times n_{i2}}$.

Now by Corollary 4.4.5, the following two conditions are equivalent:

(i) The state-space representation $(\bar{A}, \bar{B}, \bar{C}, \bar{E})$ of P, partitioned as in equation (4.4.52), satisfies the rank conditions (4.4.53) and (4.4.54) below:

$$\mathit{rank} \begin{bmatrix} sI_n - \bar{A} & -\bar{B}_2 \\ \bar{C}_1 & \bar{E}_{12} \end{bmatrix} \geq n, \quad \text{for all } s \in \mathcal{U}, \tag{4.4.53}$$

$$\mathit{rank} \begin{bmatrix} sI_n - \bar{A} & -\bar{B}_1 \\ \bar{C}_2 & \bar{E}_{21} \end{bmatrix} \geq n, \quad \text{for all } s \in \mathcal{U}. \tag{4.4.54}$$

(ii) The plant P has no decentralized fixed-eigenvalues in $\mathcal{U}$.

Furthermore, if $P \in \mathfrak{m}(\mathbb{R}_{sp}(s))$, then the equivalent conditions (i) and (ii) above are also equivalent to:

(iii) There exists a decentralized H–stabilizing compensator C_d for P .

For the special b.c.f.r. $(N_{pr} , D , N_{pl} , G)$ given in equation (4.4.52), the rank conditions (4.4.53)-(4.4.54) are the same as (4.4.44)-(4.4.45); for simplicity, we omitted the factor $\frac{1}{(s+a)}$ in equations (4.4.53) and (4.4.54) since $-a \notin \bar{\mathrm{U}}$.

If P has any decentralized fixed-eigenvalues, then these are a subset of the eigenvalues; hence, since they automatically hold for all other $s \in \mathbb{C}$, conditions (4.4.53)-(4.4.54) need to be checked only at the $\bar{\mathrm{U}}$–eigenvalues of $\bar{A}$, i.e., only for those $s \in \bar{\mathrm{U}}$ such that $\det(s I_n - \bar{A}) = 0$. Similarly, conditions (4.4.44)-(4.4.45) need to be checked only at the $\bar{\mathrm{U}}$–zeros of $\det D(s)$.

Corollary 4.4.7. (Decentralized fixed-eigenvalues of P remain in $\mathrm{S}(P , C_d)$)

Let P satisfy Assumption 4.2.1 (i). Let $s_o \in \bar{\mathrm{U}}$ be a decentralized fixed-eigenvalue of P ; then the closed-loop I/O map $H_{\overline{yu}}$ of the system $\mathrm{S}(P , C_d)$ also has a pole at $s_o \in \bar{\mathrm{U}}$ for *all dynamic* decentralized compensators C_d .

Proof

Suppose that $s_o \in \bar{\mathrm{U}}$ is a decentralized fixed-eigenvalue of P ; then by Theorem 4.4.4, statement (iv) fails at $s_o \in \bar{\mathrm{U}}$. Suppose, without loss of generality, that condition (4.4.10) fails, i.e., that

$$\mathit{rank} \begin{bmatrix} D_{p1}(s_o) \\ N_{p1}(s_o) \end{bmatrix} < n_{i1} , \tag{4.4.55}$$

where (N_p , D_p) is any r.c.f.r. of P . Equation (4.4.55) implies that

$$rank\ (\begin{bmatrix} \tilde{D}_{c1}(s_o) & \tilde{N}_{c1}(s_o) \end{bmatrix} \begin{bmatrix} D_{p1}(s_o) \\ N_{p1}(s_o) \end{bmatrix}) \le rank \begin{bmatrix} D_{p1}(s_o) \\ N_{p1}(s_o) \end{bmatrix} < n_{i1}\ , \quad (4.4.56)$$

for all $\tilde{D}_{c1}(s_o) \in \mathbb{C}^{n_{i1} \times n_{i1}}$, $\tilde{N}_{c1}(s_o) \in \mathbb{C}^{n_{i1} \times n_{o1}}$. But equation (4.4.56) implies that the characteristic determinant D_{H1}, defined in equation (4.2.19), has a zero at $s_o \in \bar{\mathcal{U}}$ since

$$rank\ D_{H1} = rank \begin{bmatrix} (\tilde{D}_{c1} D_{p1} + \tilde{N}_{c1} N_{p1})(s_o) \\ (\tilde{D}_{c2} D_{p2} + \tilde{N}_{c2} N_{p2})(s_o) \end{bmatrix}$$

$$\le rank \begin{bmatrix} (\tilde{D}_{c1} D_{p1} + \tilde{N}_{c1} N_{p1})(s_o) \end{bmatrix} + rank \begin{bmatrix} (\tilde{D}_{c2} D_{p2} + \tilde{N}_{c2} N_{p2})(s_o) \end{bmatrix} < n_{i1} + n_{i2}\ , \quad (4.4.57)$$

for all $\tilde{D}_{c1}(s_o)$, $\tilde{N}_{c1}(s_o)$, $\tilde{D}_{c2}(s_o)$, $\tilde{N}_{c2}(s_o) \in \mathfrak{m}(\mathbb{C})$; consequently, $s_o \in \bar{\mathcal{U}}$ is a pole of the closed-loop I/O map $H_{\overline{yu}}$ of the system $\mathbf{S}(P, C_d)$ since it is a zero of the characteristic determinant D_{H1} of $\mathbf{S}(P, C_d)$ for all $C_d = \tilde{D}_c^{-1} \tilde{N}_c$. □

Comment 4.4.8

(i) By Theorem 4.4.4, $s_o \in \bar{\mathcal{U}}$ is a decentralized fixed-eigenvalue of P if and only if *either*

$$rank \begin{bmatrix} D_{p1}(s_o) \\ N_{p1}(s_o) \end{bmatrix} < n_{i1} \quad (4.4.58)$$

or

$$rank \begin{bmatrix} D_{p2}(s_o) \\ N_{p2}(s_o) \end{bmatrix} < n_{i2}\ . \quad (4.4.59)$$

Now (4.4.58) and (4.4.59) cannot *both* hold at any $s_o \in \bar{\mathcal{U}}$ since (N_p, D_p) is an r.c.f.r. of P; i.e., if conditions (4.4.10) and (4.4.11) *both* failed at some $s_o \in \bar{\mathcal{U}}$, then by (4.4.58)-(4.4.59), this would imply that

$$\mathrm{rank}\begin{bmatrix} D_p(s_o) \\ N_p(s_o) \end{bmatrix} \le \mathrm{rank}\begin{bmatrix} D_{p1}(s_o) \\ N_{p1}(s_o) \end{bmatrix} + \mathrm{rank}\begin{bmatrix} D_{p2}(s_o) \\ N_{p2}(s_o) \end{bmatrix} < n_{i1} + n_{i2} . \quad (4.4.60)$$

But by Lemma 2.6.1 (i), equation (4.4.60) contradicts the fact that (N_p , D_p) is a r.c. pair.

The same comments apply to the rank conditions on a l.c.f.r. $(\tilde{D}_p , \tilde{N}_p)$, on a b.c.f.r. $(N_{pr} , D , N_{pl} , G)$ and on the state-space representation $(\bar{A} , \bar{B} , \bar{C} , \bar{E})$ of P ; i.e, conditions (4.4.12) and (4.4.13) cannot *both* fail, conditions (4.4.44) and (4.4.45) cannot *both* fail, conditions (4.4.53) and (4.4.54) cannot *both* fail at any $s_o \in \bar{\mathcal{U}}$.

(ii) By Theorem 4.4.4, if P has no decentralized fixed-eigenvalues in $\bar{\mathcal{U}}$, then conditions (4.4.6)-(4.4.7) imply that the matrix $\begin{bmatrix} D_{p1} \\ N_{p1} \end{bmatrix}$ can be put in the (Smith) form $\begin{bmatrix} I_{n_{i1}} & 0 \\ 0 & \hat{W}_1 \end{bmatrix}$ *and* at the same time, the matrix $\begin{bmatrix} D_{p2} \\ N_{p2} \end{bmatrix}$ can be put in the (Smith) form $\begin{bmatrix} 0 & I_{n_{i2}} \\ \hat{W}_2 & 0 \end{bmatrix}$, where $\hat{W}_1 \in \mathrm{H}^{n_{o1} \times n_{i2}}$ and $\hat{W}_2 \in \mathrm{H}^{n_{o2} \times n_{i1}}$; the only nonzero entries of $\hat{W}_1$ and $\hat{W}_2$ may be the ones on the diagonal; some of these diagonal terms may actually be in J (equivalently, be equal to 1). Therefore, conditions (4.4.6)-(4.4.7) imply that for $j = 1 , 2$, the first n_{ij} invariant factors of $\begin{bmatrix} D_{pj} \\ N_{pj} \end{bmatrix}$ are equal to 1 ; if the n_{ij}–th invariant factor is zero at some $s_o \in \bar{\mathcal{U}}$ for either $j = 1$ or $j = 2$, then $s_o \in \bar{\mathcal{U}}$ is a fixed-eigenvalue of P .

(iii) By Assumption 4.2.1 (i) on P , considering the b.c.f.r. $(N_{pr} , D , N_{pl} , G)$, which is partitioned as in equation (4.2.3), we can write P as

$$P = \begin{bmatrix} P_{11} & P_{12} \\ P_{21} & P_{22} \end{bmatrix} = \begin{bmatrix} N_{pr1} D^{-1} N_{pl1} + G_{11} & N_{pr1} D^{-1} N_{pl2} + G_{12} \\ N_{pr2} D^{-1} N_{pl1} + G_{21} & N_{pr2} D^{-1} N_{pl2} + G_{22} \end{bmatrix} . \quad (4.4.61)$$

If $(N_{pr1}, D, N_{pl1}, G_{11})$ is a b.c.f.r. of P_{11}, then the plant P can have no decentralized fixed-eigenvalues in $\bar{\mathcal{U}}$; similarly, if $(N_{pr2}, D, N_{pl2}, G_{22})$ is a b.c.f.r. of P_{22}, then the plant P can have no decentralized fixed-eigenvalues in $\bar{\mathcal{U}}$: To see this, note that if (N_{pr1}, D, N_{pl1}) is a b.c. triple, then by Lemma 2.6.1 (i), $rank \begin{bmatrix} D(s) \\ N_{pr1}(s) \end{bmatrix} = n$ for all $s \in \bar{\mathcal{U}}$ since (N_{pr1}, D) is an r.c. pair (this shows that condition (4.4.44) holds) *and* by Lemma 2.6.1 (ii), $rank \begin{bmatrix} D(s) & -N_{pl1}(s) \end{bmatrix} = n$ for all $s \in \bar{\mathcal{U}}$ since (D, N_{pl1}) is an l.c. pair (this shows that condition (4.4.45) holds).

In terms of the state-space representation of P as in Remark 4.4.6, this same sufficient condition for P to have no decentralized fixed-eigenvalues in $\bar{\mathcal{U}}$ is restated as follows: If $(\bar{C}_1, (sI_n - \bar{A}), \bar{B}_1, \bar{E}_{11})$ is a $\bar{\mathcal{U}}$–stabilizable and $\bar{\mathcal{U}}$–detectable state-space representation of P_{11}, then then the plant P can have no decentralized fixed-eigenvalues in $\bar{\mathcal{U}}$; the same conclusion follows if $(\bar{C}_2, (sI_n - \bar{A}), \bar{B}_2, \bar{E}_{22})$ is a $\bar{\mathcal{U}}$–stabilizable and $\bar{\mathcal{U}}$–detectable state-space representation of P_{22}.

(iv) Suppose that $s_o \in \bar{\mathcal{U}}$ is a decentralized fixed-eigenvalue of P; then by Corollary 4.4.5, either condition (4.4.44) fails or condition (4.4.45) fails. Now if (4.4.44) fails, then (N_{pr1}, D, N_{pl2}) is not a b.c. triple and hence, $P_{12} = N_{pr1} D^{-1} N_{pl2} + G_{12}$ has a hidden-mode associated with the eigenvalue $s_o \in \bar{\mathcal{U}}$ of P. If, on the other hand, (4.4.45) fails, then (N_{pr2}, D, N_{pl1}) is not a b.c. triple and hence, $P_{21} = N_{pr2} D^{-1} N_{pl1} + G_{21}$ has a hidden-mode associated with the eigenvalue $s_o \in \bar{\mathcal{U}}$ of P.

(v) Let n_{o1} be equal to n_{i1} and let n_{o2} be equal to n_{i2}. Following Comment 4.4.8 (iv) above, suppose that $s_o \in \bar{\mathcal{U}}$ is a decentralized fixed-eigenvalue of P; suppose, without loss of generality, that this decentralized fixed-eigenvalue is due to the failure of condition (4.4.44) at $s_o \in \bar{\mathcal{U}}$. Therefore,

$$rank \begin{bmatrix} D(s_o) \\ N_{pr1}(s_o) \end{bmatrix} < n \tag{4.4.62}$$

and

$$rank \begin{bmatrix} D(s_o) & -N_{pl2}(s_o) \end{bmatrix} < n . \tag{4.4.63}$$

But (4.4.62) implies that

$$rank \begin{bmatrix} D(s_o) & -N_{pl1}(s_o) \\ N_{pr1}(s_o) & G_{11}(s_o) \end{bmatrix} < n + n_{i1} \tag{4.4.64}$$

and (4.4.63) implies that

$$rank \begin{bmatrix} D(s_o) & -N_{pl2}(s_o) \\ N_{pr2}(s_o) & G_{22}(s_o) \end{bmatrix} < n + n_{i2} ; \tag{4.4.65}$$

(note that here we assume $n_{i2} = n_{o2}$). Furthermore, the failure of (4.4.44) also implies that

$$rank \begin{bmatrix} D(s_o) & -N_{pl1}(s_o) & -N_{pl2}(s_o) \\ N_{pr1}(s_o) & G_{11}(s_o) & G_{12}(s_o) \end{bmatrix} < n + n_{i1} \tag{4.4.66}$$

and therefore, by (4.4.66),

$$rank \begin{bmatrix} D(s_o) & -N_{pl1}(s_o) & -N_{pl2}(s_o) \\ N_{pr1}(s_o) & G_{11}(s_o) & G_{12}(s_o) \\ N_{pr2}(s_o) & G_{21}(s_o) & G_{22}(s_o) \end{bmatrix} < n + n_{i1} + n_{i2} . \tag{4.4.67}$$

Assuming that P has normal rank equal to n_i P_{11} has normal rank equal to $n_{i1} = n_{o1}$, P_{22} has normal rank equal to $n_{i2} = n_{o2}$, equation (4.4.64) implies that P_{11} has a *transmission-zero*, equation (4.4.65) implies that P_{22} has a *transmission-zero*, and equation (4.4.67) means that P has a *transmission-zero* at the decentralized fixed-eigenvalue $s_o \in \bar{\mathrm{U}}$. It can be shown similarly, that if the decentralized fixed-eigenvalue at $s_o \in \bar{\mathrm{U}}$ was due to the failure of condition (4.4.45) at $s_o \in \bar{\mathrm{U}}$, then again, each of P_{11}, P_{22} and P have a *transmission-zero* at the decentralized fixed-eigenvalue $s_o \in \bar{\mathrm{U}}$. □

4.4.9. Algorithm for two-channel decentralized $\mathbf{R_u}$–stabilizing compensator design

The proof of Theorem 4.4.4 (*(iv)* => *(ii)*) suggests the following algorithm to find the class of all decentralized $\mathbf{R_u}$–stabilizing compensators for a strictly proper, two-channel plant using Theorem 4.3.5. This algorithm is based on any r.c.f.r. (N_p , D_p) of P ; similar algorithms can be obtained based on any l.c.f.r. or b.c.f.r. of P as well.

Given: a plant $P \in \mathbb{R}_{sp}(s)^{n_o \times n_i}$ which satisfies Assumption 4.2.1 (i).

Find: an r.c.f.r. (N_p , D_p) of P ; partition D_p and N_p as in equation (4.2.1).

Check: that D_{p1}, N_{p1} satisfy the rank condition (4.4.10) and that D_{p2}, N_{p2} satisfy the rank condition (4.4.11). If either one of these conditions fails, then stop; there is no decentralized $\mathbf{R_u}$–stabilizing compensator for this plant.

Step 1: Find $\mathbf{R_u}$–unimodular matrices L_1 and $R_1 \in \mathfrak{m}(\mathbf{R_u})$ such that

$$L_1 \begin{bmatrix} D_{p1} \\ N_{p1} \end{bmatrix} R_1 = \begin{bmatrix} I_{n_{i1}} & 0 \\ 0 & \hat{N}_{12} \end{bmatrix} , \tag{4.4.68}$$

where $\hat{N}_{12} \in \mathbf{R_u}^{n_{o1} \times n_{i2}}$ is some $\mathbf{R_u}$–stable matrix.

Step 2: Find an $\mathbf{R_u}$–unimodular matrix $L_2 \in \mathfrak{m}(\mathbf{R_u})$ such that

$$L_2 (\begin{bmatrix} D_{p2} \\ N_{p2} \end{bmatrix} R_1) = \begin{bmatrix} -\hat{D}_{21} & \hat{D}_{22} \\ \hat{N}_{21} & 0 \end{bmatrix} , \tag{4.4.69}$$

where $\hat{D}_{22} \in \mathbf{R_u}^{n_{i2} \times n_{i2}}$, $\hat{D}_{21} \in \mathbf{R_u}^{n_{i2} \times n_{i1}}$, $\hat{N}_{21} \in \mathbf{R_u}^{n_{o2} \times n_{i1}}$ are some $\mathbf{R_u}$–stable matrices and $\hat{D}_{21}$, $\hat{D}_{22}$ also satisfy

$$(\hat{D}_{22} , \hat{D}_{21}) \quad \text{is an l.c. pair ;} \tag{4.4.70}$$

equivalently, $rank \begin{bmatrix} \hat{D}_{21} & \hat{D}_{22} \end{bmatrix} = n_{i2}$, for all $s \in \bar{\mathcal{U}}$.

Step 3: Find a generalized Bezout identity for the l.c. pair $(\hat{D}_{22}, \hat{D}_{21})$; i.e., find matrices $V_{2l}, U_{2l}, X_2, Y_2, U_2, V_2 \in \mathrm{m}(\mathbf{R_u})$ such that

$$\begin{bmatrix} V_2 & U_2 \\ -\hat{D}_{21} & \hat{D}_{22} \end{bmatrix} \begin{bmatrix} Y_2 & -U_{2l} \\ X_2 & V_{2l} \end{bmatrix} = \begin{bmatrix} I_{n_{i1}} & 0 \\ 0 & I_{n_{i2}} \end{bmatrix} . \tag{4.4.71}$$

Step 4: Find a generalized Bezout identity for the r.c. pair $(\hat{N}_{12}, \hat{D}_{22})$; i.e., find matrices $V_{2r}, U_{2r}, \tilde{X}_2, \tilde{Y}_2, \tilde{U}_2, \tilde{V}_2 \in \mathrm{m}(\mathbf{R_u})$ such that

$$\begin{bmatrix} V_{2r} & U_{2r} \\ -\tilde{X}_2 & \tilde{Y}_2 \end{bmatrix} \begin{bmatrix} \hat{D}_{22} & -\tilde{U}_2 \\ \hat{N}_{12} & \tilde{V}_2 \end{bmatrix} = \begin{bmatrix} I_{n_{i2}} & 0 \\ 0 & I_{n_{o1}} \end{bmatrix} . \tag{4.4.72}$$

Step 5: Define

$$R_2 := \begin{bmatrix} Y_2 & -U_{2l} \\ X_2 & V_{2l} \end{bmatrix} \in \mathbf{R_u}^{n_i \times n_i} ;$$

$$R := R_1 R_2 \begin{bmatrix} I_{n_{i1}} & U_2 V_{2r} \\ 0 & I_{n_{i2}} \end{bmatrix} \in \mathbf{R_u}^{n_i \times n_i} . \tag{4.4.73}$$

Define

$$E_1 := \begin{bmatrix} V_2 + U_2 V_{2r} \hat{D}_{21} & U_2 U_{2r} \\ -\tilde{X}_2 \hat{D}_{21} & \tilde{Y}_2 \end{bmatrix} L_1 \in \mathbf{R_u}^{(n_{i1}+n_{o1}) \times (n_{i1}+n_{o1})} ; \tag{4.4.74}$$

$$E_2 := \begin{bmatrix} I_{n_{i2}} & 0 \\ \hat{N}_{21} U_{2l} \tilde{U}_2 \tilde{X}_2 & I_{n_{o2}} \end{bmatrix} L_2 \in \mathbf{R_u}^{(n_{i2}+n_{o2}) \times (n_{i2}+n_{o2})} . \tag{4.4.75}$$

Finally define

$$W_{12} := \tilde{X}_2 \in \mathbf{R_U}^{n_{o1} \times n_{i2}} \quad \text{and} \quad W_{21} := \hat{N}_{21} Y_2 \in \mathbf{R_U}^{n_{o2} \times n_{i1}} . \tag{4.4.76}$$

Step 6: Find matrices $Q_{11} \in \mathbf{R_U}^{n_{i1} \times n_{i1}}$, $Q_1 \in \mathbf{R_U}^{n_{i1} \times n_{o1}}$, $Q_2 \in \mathbf{R_U}^{n_{i2} \times n_{o2}}$, $Q_{22} \in \mathbf{R_U}^{n_{i2} \times n_{i2}}$ such that

$$\begin{bmatrix} Q_{11} & Q_1 W_{12} \\ Q_2 W_{21} & Q_{22} \end{bmatrix} \text{ is } \mathbf{R_U}\text{–unimodular} . \tag{4.4.77}$$

Step 7: Let

$$\begin{bmatrix} \tilde{D}_{c1} & \tilde{N}_{c1} \end{bmatrix} = \begin{bmatrix} Q_{11} & Q_1 \end{bmatrix} E_1 \tag{4.4.78}$$

and let

$$\begin{bmatrix} \tilde{D}_{c2} & \tilde{N}_{c2} \end{bmatrix} = \begin{bmatrix} Q_{22} & Q_2 \end{bmatrix} E_2 . \tag{4.4.79}$$

Let $C_d = \begin{bmatrix} C_1 & 0 \\ 0 & C_2 \end{bmatrix} = \begin{bmatrix} \tilde{D}_{c1}^{-1} \tilde{N}_{c1} & 0 \\ 0 & \tilde{D}_{c2}^{-1} \tilde{N}_{c2} \end{bmatrix}$, where $\tilde{D}_{c1}$, $\tilde{N}_{c1}$ are given by (4.4.78) and $\tilde{D}_{c2}$, $\tilde{N}_{c2}$ are given by (4.4.79). □

Example 4.4.10

We now follow the steps of Algorithm 4.4.9 to find a decentralized $\mathbf{R_U}$–stabilizing compensator for the plant

$$P = \begin{bmatrix} 0 & \dfrac{1}{s-2} \\ \dfrac{1}{s-1} & \dfrac{-(s+1)}{(s-1)(s-2)} \end{bmatrix} , \tag{4.4.80}$$

where $n_{i1} = n_{i2} = 1$. Let $\bar{\mathbf{U}}$ be the set of all $s \in \mathbf{C}$ whose real part is greater than or equal to -1. An r.c.f.r. (N_p, D_p) of P is then given by

$$N_p = \begin{bmatrix} \frac{1}{s+4} & 0 \\ 0 & \frac{1}{s+4} \end{bmatrix}, \quad D_p = \begin{bmatrix} \frac{s+1}{s+4} & \frac{s-1}{s+4} \\ \frac{s-2}{s+4} & 0 \end{bmatrix}. \tag{4.4.81}$$

Now the rank conditions (4.4.10)-(4.4.11) hold since

$$rank \begin{bmatrix} D_{p1}(s) \\ N_{p1}(s) \end{bmatrix} = rank \begin{bmatrix} \frac{s+1}{s+4} & \frac{s-1}{s+4} \\ \frac{1}{s+4} & 0 \end{bmatrix} \geq n_{i1} = 1 , \quad \text{and}$$

$$rank \begin{bmatrix} D_{p2}(s) \\ N_{p2}(s) \end{bmatrix} = rank \begin{bmatrix} \frac{s-2}{s+4} & 0 \\ 0 & \frac{1}{s+4} \end{bmatrix} \geq n_{i2} = 1 , \quad \text{for all } s \in \bar{\mathcal{U}} .$$

Step 1: One choice for the $\mathbf{R_U}$–unimodular matrices L_1 and R_1 in equation (4.4.68) is:

$$L_1 = \begin{bmatrix} 1 & 3 \\ \frac{-1}{s+4} & \frac{s+1}{s+4} \end{bmatrix}, \quad R_1 = \begin{bmatrix} 1 & \frac{s-1}{s+4} \\ 0 & -1 \end{bmatrix}, \tag{4.4.82}$$

and hence, $\hat{N}_{12} = \dfrac{s-1}{(s+4)^2}$.

Step 2: One choice for the $\mathbf{R_U}$–unimodular matrix $L_2 \in \mathbf{m}(\mathbf{R_U})$ in equation (4.4.69) is:

$$L_2 = \begin{bmatrix} 1 & \frac{-(11s+14)}{s+4} \\ \frac{1}{s+4} & \frac{(s-1)(s-2)}{(s+4)^2} \end{bmatrix}. \tag{4.4.83}$$

With L_2 as in equation (4.4.83), we get

$$\begin{bmatrix} -\hat{D}_{21} & \hat{D}_{22} \\ \hat{N}_{21} & 0 \end{bmatrix} = \begin{bmatrix} \dfrac{s-2}{s+4} & 1 \\ \dfrac{s-2}{(s+4)^2} & 0 \end{bmatrix} . \qquad (4.4.84)$$

From (4.4.84), the pair $(\hat{D}_{22}, \hat{D}_{21}) = (1, \dfrac{-(s-2)}{s+4})$ is coprime.

Step 3: A generalized Bezout identity as in (4.4.71) for the l.c. pair $(\hat{D}_{22}, \hat{D}_{21})$ is given by

$$\begin{bmatrix} 1 & 0 \\ \dfrac{s-2}{s+4} & 1 \end{bmatrix} \begin{bmatrix} 1 & 0 \\ \dfrac{-(s-2)}{s+4} & 1 \end{bmatrix} = \begin{bmatrix} 1 & 0 \\ 0 & 1 \end{bmatrix} . \qquad (4.4.85)$$

Step 4: A generalized Bezout identity as in (4.4.72) for the r.c. pair $(\hat{N}_{12}, \hat{D}_{22})$ is given by

$$\begin{bmatrix} 1 & 0 \\ \dfrac{-(s-1)}{(s+4)^2} & 1 \end{bmatrix} \begin{bmatrix} 1 & 0 \\ \dfrac{s-1}{(s+4)^2} & 1 \end{bmatrix} = \begin{bmatrix} 1 & 0 \\ 0 & 1 \end{bmatrix} . \qquad (4.4.86)$$

Step 5: By equation (4.4.73),

$$R = \begin{bmatrix} \dfrac{11s+14}{(s+4)^2} & \dfrac{s-1}{s+4} \\ \dfrac{s-2}{s+4} & -1 \end{bmatrix} . \qquad (4.4.87)$$

By equations (4.4.74)-(4.4.75),

$$E_1 = \begin{bmatrix} 1 & 3 \\ \dfrac{-(11s+14)}{(s+4)^3} & \dfrac{s^3+12s^2+15s+22}{(s+4)^3} \end{bmatrix} , \qquad (4.4.88)$$

$$E_2 = \begin{bmatrix} 1 & \dfrac{-(11s+14)}{s+4} \\ \dfrac{1}{s+4} & \dfrac{(s-1)(s-2)}{(s+4)^2} \end{bmatrix} = L_2 \,. \tag{4.4.89}$$

By equation (4.4.76),

$$W_{12} = \frac{s-1}{(s+4)^2} \quad , \quad W_{21} = \frac{s-2}{(s+4)^2} \,. \tag{4.4.90}$$

Step 6: Choose $Q_{11} = 1$, $Q_{22} = 1$, $Q_1 = 0$ and $Q_2 = 0$. Condition (4.4.77) is satisfied with this choice of Q_{11}, Q_1, Q_2 and $Q_{22} \in \mathbf{R_u}$ since

$$\begin{bmatrix} Q_{11} & Q_1 W_{12} \\ Q_2 W_{21} & Q_{22} \end{bmatrix} = \begin{bmatrix} 1 & 0 \\ 0 & 1 \end{bmatrix}.$$

Step 7: By equation (4.4.78),

$$\begin{bmatrix} \tilde{D}_{c1} & \tilde{N}_{c1} \end{bmatrix} = \begin{bmatrix} 1 & 0 \end{bmatrix} \begin{bmatrix} 1 & 3 \\ \dfrac{-(11s+14)}{(s+4)^3} & \dfrac{s^3+12s^2+15s+22}{(s+4)^3} \end{bmatrix}$$

$$= \begin{bmatrix} 1 & 3 \end{bmatrix}, \tag{4.4.91}$$

and by equation (4.4.79),

$$\begin{bmatrix} \tilde{D}_{c2} & \tilde{N}_{c2} \end{bmatrix} = \begin{bmatrix} 1 & 0 \end{bmatrix} \begin{bmatrix} 1 & \dfrac{-(11s+14)}{s+4} \\ \dfrac{1}{s+4} & \dfrac{(s-1)(s-2)}{(s+4)^2} \end{bmatrix}$$

$$= \begin{bmatrix} 1 & \dfrac{-(11s+14)}{s+4} \end{bmatrix}. \tag{4.4.92}$$

Finally by equations (4.4.91) and (4.4.92),

$$C_d = \begin{bmatrix} \tilde{D}_{c1}^{-1} \tilde{N}_{c1} & 0 \\ 0 & \tilde{D}_{c2}^{-1} \tilde{N}_{c2} \end{bmatrix} = \begin{bmatrix} 3 & 0 \\ 0 & \dfrac{-(11s+14)}{s+4} \end{bmatrix} . \tag{4.4.93}$$

The decentralized $\mathbf{R_u}$–stabilizing compensator C_d in equation (4.4.93) is itself $\mathbf{R_u}$–stable; Note that this is a not always possible, i.e., the plant need not be $\mathbf{R_u}$–stabilizable by an $\mathbf{R_u}$–stable compensator in general. For example, if we chose $Q_{11} = 15$, $Q_{22} = 1$, $Q_1 = -1$, $Q_2 = -5$, then condition (4.4.77) is still satisfied since

$$\begin{bmatrix} Q_{11} & Q_1 W_{12} \\ Q_2 W_{21} & Q_{22} \end{bmatrix} = \begin{bmatrix} 15 & \dfrac{-(s-1)}{(s+4)^2} \\ \dfrac{-5(s-2)}{(s+4)^2} & 1 \end{bmatrix}$$

is $\mathbf{R_u}$–unimodular. Repeating Step 7, we obtain another decentralized $\mathbf{R_u}$–stabilizing compensator for the plant in (4.4.80):

$$C_d = \begin{bmatrix} \dfrac{44s^3+528s^2+2145s+2858}{15s^3+180s^2+731s+974} & 0 \\ 0 & \dfrac{-(16s^2+43s+66)}{(s+4)(s-1)} \end{bmatrix} . \tag{4.4.94}$$

The decentralized $\mathbf{R_u}$–stabilizing compensator C_d in (4.4.94) is *not* $\mathbf{R_u}$–stable. □

4.5 MULTI-CHANNEL DECENTRALIZED CONTROL SYSTEMS

In this section we extend the results of Sections 4.3 and 4.4 to m-channel decentralized feedback control systems, where $m > 2$.

We consider the linear, time-invariant, m-channel decentralized feedback system $\mathbf{S}(P, C_d)_m$ shown in Figure 4.7, where $P : \begin{bmatrix} e_1 \\ \vdots \\ e_m \end{bmatrix} \mapsto \begin{bmatrix} y_1 \\ \vdots \\ y_m \end{bmatrix}$ represents the plant and

$C_d : \begin{bmatrix} e_1' \\ \vdots \\ e_m' \end{bmatrix} \mapsto \begin{bmatrix} y_1' \\ \vdots \\ y_m' \end{bmatrix}$ represents the compensator.

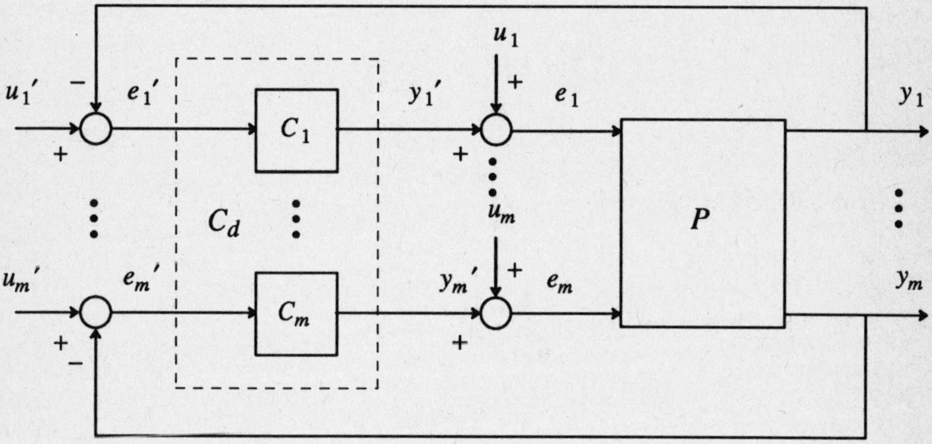

Figure 4.7. The m-channel decentralized control system $\mathbf{S}(P, C_d)_m$.

The externally applied inputs are denoted by $\bar{u} := \begin{bmatrix} u_1 \\ \vdots \\ u_m \\ u_1' \\ \vdots \\ u_m' \end{bmatrix}$, the plant and the compensator outputs are denoted by $\bar{y} := \begin{bmatrix} y_1 \\ \vdots \\ y_m \\ y_1' \\ \vdots \\ y_m' \end{bmatrix}$; the closed-loop input-output map of $\mathbf{S}(P, C_d)$ is denoted by $H_{\bar{y}\bar{u}} : \bar{u} \mapsto \bar{y}$. We extend Assumption 4.2.1 to m-channels as follows:

4.5.1. Assumptions on $\mathbf{S}(P, C_d)_m$

(i) The m-channel plant $P \in \mathbf{G}^{n_o \times n_i}$, where

$$n_o = n_{o1} + n_{o2} + \cdots + n_{om} \quad , \quad n_i = n_{i1} + n_{i2} + \cdots + n_{im} \; .$$

(ii) The decentralized compensator $C_d \in \mathbf{G}^{n_i \times n_o}$, where

$$C_d = \begin{bmatrix} C_1 & 0 & \cdots & 0 \\ 0 & C_2 & \cdots & 0 \\ \vdots & \vdots & \vdots & \vdots \\ 0 & 0 & \cdots & C_m \end{bmatrix} \quad \text{and for } j = 1, \cdots, m \; , \; C_j \in \mathbf{G}^{n_{ij} \times n_{oj}} \; .$$

(iii) The system $\mathbf{S}(P, C_d)_m$ is well-posed; equivalently, the closed-loop input-output map $H_{\bar{y}\bar{u}} : \bar{u} \mapsto \bar{y}$ is in $\mathrm{m}(\mathbf{G})$. □

Note that whenever P satisfies Assumption 4.5.1 (i), it has an r.c.f.r., denoted by (N_p, D_p), an l.c.f.r., denoted by $(\tilde{D}_p, \tilde{N}_p)$ and a b.c.f.r., denoted by (N_{pr}, D, N_{pl}, G), where the numerator and the denominator matrices can be partitioned as follows: In the r.c.f.r. (N_p, D_p),

$$N_p =: \begin{bmatrix} N_{p1} \\ N_{p2} \\ \vdots \\ N_{pm} \end{bmatrix} \in \mathbf{H}^{n_o \times n_i} \; , \; D_p =: \begin{bmatrix} D_{p1} \\ D_{p2} \\ \vdots \\ D_{pm} \end{bmatrix} \in \mathbf{H}^{n_i \times n_i} \; , \qquad (4.5.1)$$

where, for $j = 1, \cdots, m$, $N_{pj} \in \mathbf{H}^{n_{oj} \times n_i}$, $D_{pj} \in \mathbf{H}^{n_{ij} \times n_i}$.

In the l.c.f.r. $(\tilde{D}_p, \tilde{N}_p)$,

$$\tilde{D}_p =: \begin{bmatrix} \tilde{D}_{p1} & \tilde{D}_{p2} & \cdots & \tilde{D}_{pm} \end{bmatrix} \in \mathrm{H}^{n_o \times n_o} \quad ,$$

$$\tilde{N}_p =: \begin{bmatrix} \tilde{N}_{p1} & \tilde{N}_{p2} & \cdots & \tilde{N}_{pm} \end{bmatrix} \in \mathrm{H}^{n_o \times n_i} \quad , \tag{4.5.2}$$

where, for $j = 1, \cdots, m$, $\tilde{D}_{pj} \in \mathrm{H}^{n_o \times n_{oj}}$, $\tilde{N}_{pj} \in \mathrm{H}^{n_o \times n_{ij}}$.

In the b.c.f.r. (N_{pr}, D, N_{pl}, G),

$$N_{pr} =: \begin{bmatrix} N_{pr1} \\ N_{pr2} \\ \vdots \\ N_{prm} \end{bmatrix} \in \mathrm{H}^{n_o \times n} \quad , \quad N_{pl} =: \begin{bmatrix} N_{pl1} & N_{pl2} & \cdots & N_{plm} \end{bmatrix} \in \mathrm{H}^{n \times n_i} \; ,$$

$$G =: \begin{bmatrix} G_{11} & \cdots & G_{1m} \\ G_{21} & \cdots & G_{2m} \\ \vdots & & \vdots \\ G_{m1} & \cdots & G_{mm} \end{bmatrix} , \tag{4.5.3}$$

where, for $j = 1, \cdots, m$, $k = 1, \cdots, m$, $N_{prj} \in \mathrm{H}^{n_{oj} \times n}$, $N_{plj} \in \mathrm{H}^{n \times n_{ij}}$, $G_{jk} \in \mathrm{H}^{n_{oj} \times n_{ik}}$; $D \in \mathrm{H}^{n \times n}$.

If C_d satisfies Assumption 4.5.1 **(ii)**, then C_d has an l.c.f.r., denoted by $(\tilde{D}_c, \tilde{N}_c)$ and an r.c.f.r., denoted by (N_c, D_c), where $\tilde{D}_c \in \mathrm{H}^{n_i \times n_i}$, $\tilde{N}_c \in \mathrm{H}^{n_i \times n_o}$, $N_c \in \mathrm{H}^{n_i \times n_o}$, $D_c \in \mathrm{H}^{n_o \times n_o}$. Let

$$\tilde{D}_c = \begin{bmatrix} \tilde{D}_{c1} & 0 & \cdots & 0 \\ 0 & \tilde{D}_{c2} & \cdots & 0 \\ \vdots & \vdots & & \vdots \\ 0 & 0 & \cdots & \tilde{D}_{cm} \end{bmatrix} , \quad \tilde{N}_c = \begin{bmatrix} \tilde{N}_{c1} & 0 & \cdots & 0 \\ 0 & \tilde{N}_{c2} & \cdots & 0 \\ \vdots & \vdots & & \vdots \\ 0 & 0 & \cdots & \tilde{N}_{cm} \end{bmatrix} ; \tag{4.5.4}$$

note that for $j = 1, \cdots, m$, $(\tilde{D}_c, \tilde{N}_c)$ is an l.c.f.r. of C_d if and only if $(\tilde{D}_{cj}, \tilde{N}_{cj})$ is an l.c.f.r. of C_j, where $\tilde{D}_{cj} \in \mathrm{H}^{n_{ij} \times n_{ij}}$, $\tilde{N}_{cj} \in \mathrm{H}^{n_{ij} \times n_{oj}}$. Let

$$D_c = \begin{bmatrix} D_{c1} & 0 & \cdots & 0 \\ 0 & D_{c2} & \cdots & 0 \\ \vdots & \vdots & & \vdots \\ 0 & 0 & \cdots & D_{cm} \end{bmatrix}, \quad N_c = \begin{bmatrix} N_{c1} & 0 & \cdots & 0 \\ 0 & N_{c2} & \cdots & 0 \\ \vdots & \vdots & & \vdots \\ 0 & 0 & \cdots & N_{cm} \end{bmatrix}; \tag{4.5.5}$$

note that for $j = 1, \cdots, m$, (N_c, D_c) is an r.c.f.r. of C_d if and only if (N_{cj}, D_{cj}) is an r.c.f.r. of C_j, where $D_{cj} \in \mathrm{H}^{n_{oj} \times n_{oj}}$, $N_{cj} \in \mathrm{H}^{n_{ij} \times n_{oj}}$. □

4.5.2. Analysis (Descriptions of $\mathbf{S}(P, C_d)_m$ using coprime factorizations)

We only analyze the m-channel decentralized system as in Analysis 4.2.4 **(i)**; the other cases of Section 4.2 are also easy to extend.

Assumptions 4.5.1 hold throughout this analysis.

Let (N_p, D_p) be any r.c.f.r. of $P \in \mathfrak{m}(\mathrm{G})$, partitioned as in equation (4.5.1) and let $(\tilde{D}_c, \tilde{N}_c)$ be any l.c.f.r. of $C_d \in \mathfrak{m}(\mathrm{G})$, partitioned as in equation (4.5.4). The m-channel system $\mathbf{S}(P, C_d)_m$ is then described by equations (4.5.6)-(4.5.7) below:

$$\begin{bmatrix} \tilde{D}_{c1} D_{p1} + \tilde{N}_{c1} N_{p1} \\ \vdots \\ \tilde{D}_{cm} D_{pm} + \tilde{N}_{cm} N_{pm} \end{bmatrix} \xi_p = \begin{bmatrix} \tilde{D}_{c1} & \cdots & 0 & \tilde{N}_{c1} & \cdots & 0 \\ \vdots & & \vdots & \vdots & & \\ 0 & \cdots & \tilde{D}_{cm} & 0 & \cdots & \tilde{N}_{cm} \end{bmatrix} \begin{bmatrix} u_1 \\ \vdots \\ u_m \\ u_1' \\ \vdots \\ u_m' \end{bmatrix}, \tag{4.5.6}$$

$$\begin{bmatrix} N_{p1} \\ \vdots \\ N_{pm} \\ D_{p1} \\ \vdots \\ D_{pm} \end{bmatrix} \xi_p = \begin{bmatrix} y_1 \\ \vdots \\ y_m \\ y_1' \\ \vdots \\ y_m' \end{bmatrix} - \begin{bmatrix} 0 & \cdots & 0 & 0 & \cdots & 0 \\ \vdots & & \vdots & \vdots & & \vdots \\ 0 & \cdots & 0 & 0 & \cdots & 0 \\ -I_{n_{i1}} & \cdots & 0 & 0 & \cdots & 0 \\ \vdots & & \vdots & \vdots & & \vdots \\ 0 & \cdots & -I_{n_{im}} & 0 & \cdots & 0 \end{bmatrix} \begin{bmatrix} u_1 \\ \vdots \\ u_m \\ u_1' \\ \vdots \\ u_m' \end{bmatrix}. \tag{4.5.7}$$

□

Theorem 4.5.3 and Theorem 4.5.4 below are obvious extensions to the m-channel case of Theorem 4.2.5 and Theorem 4.3.3, respectively; we state them without proof:

Theorem 4.5.3. (H–stability of $\mathbf{S}(P, C_d)_m$)

Let Assumptions 4.5.1 (i) and (ii) hold; let (N_p, D_p) be any r.c.f.r. over $\mathfrak{m}(\mathrm{H})$ of $P \in \mathfrak{m}(\mathrm{G})$, partitioned as in equations (4.5.1); let $(\tilde{D}_c, \tilde{N}_c)$ be any l.c.f.r. over $\mathfrak{m}(\mathrm{H})$ of $C \in \mathfrak{m}(\mathrm{G})$, partitioned as in equations (4.5.4). Under these assumptions, $\mathbf{S}(P, C_d)_m$ is H–stable if and only if the matrix D_{H1} in equation (4.5.8) is H–unimodular.

$$D_{H1} := \begin{bmatrix} \tilde{D}_c D_p + \tilde{N}_c N_p \end{bmatrix} = \begin{bmatrix} \tilde{D}_{c1} D_{p1} + \tilde{N}_{c1} N_{p1} \\ \vdots \\ \tilde{D}_{cm} D_{pm} + \tilde{N}_{cm} N_{pm} \end{bmatrix}$$

$$= \begin{bmatrix} \tilde{D}_{c1} & \tilde{N}_{c1} & \cdots & 0 & 0 \\ \vdots & \vdots & & \vdots & \vdots \\ 0 & 0 & \cdots & \tilde{D}_{cm} & \tilde{N}_{cm} \end{bmatrix} \begin{bmatrix} D_{p1} \\ N_{p1} \\ \vdots \\ D_{pm} \\ N_{pm} \end{bmatrix} . \qquad (4.5.8)$$

Theorem 4.5.4. (Conditions on $P = N_p D_p^{-1} = \tilde{D}_p^{-1} \tilde{N}_p$ for decentralized H–stabilizability)

Let P satisfy Assumption 4.5.1 (i); furthermore, let $P \in \mathfrak{m}(\mathrm{G}_s)$; then the following three conditions are equivalent:

(i) There exists a decentralized H–stabilizing compensator C_d for P ;

(ii) Any r.c.f.r. (N_p, D_p) of P, partitioned as in (4.5.1), satisfies condition (4.5.9) below:

$$\begin{bmatrix} E_1 \begin{bmatrix} D_{p1} \\ N_{p1} \end{bmatrix} \\ E_2 \begin{bmatrix} D_{p2} \\ N_{p2} \end{bmatrix} \\ \vdots \\ E_m \begin{bmatrix} D_{pm} \\ N_{pm} \end{bmatrix} \end{bmatrix} R = \begin{bmatrix} \begin{bmatrix} I_{n_{i1}} & 0 & 0 & \cdots & 0 \\ 0 & W_{12} & W_{13} & \cdots & W_{1m} \end{bmatrix} \\ \begin{bmatrix} 0 & I_{n_{i2}} & 0 & \cdots & 0 \\ W_{21} & 0 & W_{23} & \cdots & W_{2m} \end{bmatrix} \\ \vdots \\ \begin{bmatrix} 0 & 0 & 0 & \cdots & I_{n_{im}} \\ W_{m1} & W_{m2} & W_{m3} & \cdots & 0 \end{bmatrix} \end{bmatrix} , \qquad (4.5.9)$$

(iii) Any l.c.f.r. $(\tilde{D}_p, \tilde{N}_p)$ of P, partitioned as in (4.5.2), satisfies condition (4.5.10) below:

$$L \left[\left[-\tilde{N}_{p1} \quad \tilde{D}_{p1} \right] E_1^{-1} \quad \left[-\tilde{N}_{p2} \quad \tilde{D}_{p2} \right] E_2^{-1} \quad \cdots \quad \left[-\tilde{N}_{pm} \quad \tilde{D}_{pm} \right] E_m^{-1} \right]$$

$$= \left[\begin{bmatrix} 0 & I_{n_{o1}} \\ -W_{21} & 0 \\ W_{31} & 0 \\ \vdots & \vdots \\ -W_{m1} & 0 \end{bmatrix} \begin{bmatrix} -W_{12} & 0 \\ 0 & I_{n_{o2}} \\ -W_{32} & 0 \\ \vdots & \vdots \\ -W_{m2} & 0 \end{bmatrix} \cdots \begin{bmatrix} -W_{1m} & 0 \\ -W_{2m} & 0 \\ -W_{3m} & 0 \\ \vdots & \vdots \\ 0 & I_{n_{om}} \end{bmatrix} \right], \quad (4.5.10)$$

where, for $j = 1, \cdots, m$, $E_j \in \mathrm{H}^{(n_{ij}+n_{oj}) \times (n_{ij}+n_{oj})}$ is H–unimodular, $R \in \mathrm{H}^{n_i \times n_i}$ is H–unimodular and $L \in \mathrm{H}^{n_o \times n_o}$ is H–unimodular; the matrices $W_{jk} \in \mathrm{H}^{n_{oj} \times n_{ik}}$ are H–stable for $k = 1, \cdots, m$ (note that $W_{jk} = 0$ when $k = j$). □

We now extend the class of all decentralized H–stabilizing compensators in Theorem 4.3.5 to m-channels: The matrix T_m in equations (4.5.13)-(4.5.14) below is defined as

$$T_m := \begin{bmatrix} Q_{11} & Q_1 W_{12} & Q_1 W_{13} & \cdots & Q_1 W_{1m} \\ Q_2 W_{21} & Q_{22} & Q_2 W_{23} & \cdots & Q_2 W_{2m} \\ Q_3 W_{31} & Q_3 W_{32} & Q_{33} & \cdots & Q_3 W_{3m} \\ \vdots & \vdots & \vdots & & \vdots \\ Q_m W_{m1} & Q_m W_{m2} & Q_m W_{m3} & \cdots & Q_{mm} \end{bmatrix} = \begin{bmatrix} Q_{11} & 0 & 0 & \cdots & 0 \\ 0 & Q_{22} & 0 & \cdots & 0 \\ 0 & 0 & Q_{33} & \cdots & 0 \\ \vdots & \vdots & \vdots & & \vdots \\ 0 & 0 & 0 & \cdots & Q_{mm} \end{bmatrix}$$

$$+ \begin{bmatrix} Q_1 & 0 & 0 & \cdots & 0 \\ 0 & Q_2 & 0 & \cdots & 0 \\ 0 & 0 & Q_3 & \cdots & 0 \\ \vdots & \vdots & \vdots & & \vdots \\ 0 & 0 & 0 & \cdots & Q_m \end{bmatrix} \begin{bmatrix} 0 & W_{12} & W_{13} & \cdots & W_{1m} \\ W_{21} & 0 & W_{23} & \cdots & W_{2m} \\ W_{31} & W_{32} & 0 & \cdots & W_{3m} \\ \vdots & \vdots & \vdots & & \vdots \\ W_{m1} & W_{m2} & W_{m3} & \cdots & 0 \end{bmatrix}; \quad (4.5.11)$$

the matrix $\hat{T}_m$ is defined similarly by replacing each Q_j with $\hat{Q}_j$ and Q_{jj} with $\hat{Q}_{jj}$ for $j = 1, \cdots, m$.

The following generalized Bezout identity, which is an extension of equation (4.3.28) to m-channels, is useful to prove Theorem 4.5.5 below; the details of the proof can easily be worked out following the same steps as in the proof of Theorem 4.3.5.

$$
\begin{bmatrix}
R\begin{bmatrix} I_{n_i 1} & 0 \\ 0 & 0 \\ 0 & 0 \\ \vdots & \vdots \\ 0 & 0 \end{bmatrix} E_1 &
R\begin{bmatrix} 0 & 0 \\ I_{n_i 2} & 0 \\ 0 & 0 \\ \vdots & \vdots \\ 0 & 0 \end{bmatrix} E_2 & \cdots &
R\begin{bmatrix} 0 & 0 \\ 0 & 0 \\ 0 & 0 \\ \vdots & \vdots \\ I_{n_{im}} & 0 \end{bmatrix} E_m \\
L^{-1}\begin{bmatrix} 0 & I_{n_o 1} \\ -W_{21} & 0 \\ -W_{31} & 0 \\ \vdots & \vdots \\ -W_{m1} & 0 \end{bmatrix} E_1 &
L^{-1}\begin{bmatrix} -W_{12} & 0 \\ 0 & I_{n_o 2} \\ -W_{32} & 0 \\ \vdots & \vdots \\ -W_{m2} & 0 \end{bmatrix} E_2 & \cdots &
L^{-1}\begin{bmatrix} -W_{1m} & 0 \\ -W_{2m} & 0 \\ -W_{3m} & 0 \\ \vdots & \vdots \\ 0 & I_{n_{om}} \end{bmatrix} E_m
\end{bmatrix}
$$

$$
\cdot
\begin{bmatrix}
E_1^{-1}\begin{bmatrix} I_{n_i 1} & 0 & 0 & \cdots & 0 \\ 0 & W_{12} & W_{13} & \cdots & W_{1m} \end{bmatrix} R^{-1} &
E_1^{-1}\begin{bmatrix} 0 & 0 & \cdots & 0 \\ I_{n_o 1} & 0 & \cdots & 0 \end{bmatrix} L \\
E_2^{-1}\begin{bmatrix} 0 & I_{n_i 2} & 0 & \cdots & 0 \\ W_{21} & 0 & W_{23} & \cdots & W_{2m} \end{bmatrix} R^{-1} &
E_2^{-1}\begin{bmatrix} 0 & 0 & \cdots & 0 \\ 0 & I_{n_o 2} & \cdots & 0 \end{bmatrix} L \\
\vdots & \vdots \\
E_m^{-1}\begin{bmatrix} 0 & 0 & 0 & \cdots & I_{n_{im}} \\ W_{m1} & W_{m2} & W_{m3} & \cdots & 0 \end{bmatrix} R^{-1} &
E_m^{-1}\begin{bmatrix} 0 & 0 & \cdots & 0 \\ 0 & 0 & \cdots & I_{n_{om}} \end{bmatrix} L
\end{bmatrix}
= I_{n_i + n_o} \, . \tag{4.5.12}
$$

Theorem 4.5.5. (Class of all decentralized H–stabilizing compensators in $\mathrm{S}(P, C_d)_m$)

Let P satisfy Assumption 4.5.1 (i); furthermore let $P \in \mathrm{m}(\mathrm{G}_s)$; let any r.c.f.r. (N_p, D_p) of P satisfy condition (4.5.9) and equivalently, let any l.c.f.r. $(\tilde{D}_p, \tilde{N}_p)$ of P satisfy condition (4.5.10) of Theorem 4.5.4. Under these assumptions, the set $S_d(P)$ of *all* decentralized H–stabilizing compensators for P is given by

$$S_d(P) = \left[C_1 \ \cdots \ C_m \right] = diag \left[\tilde{D}_{c1}^{-1} \tilde{N}_{c1} \ \cdots \ \tilde{D}_{cm}^{-1} \tilde{N}_{cm} \right] \ \Big|$$

$$\text{for } j = 1, \cdots, m, \left[\tilde{D}_{cj} \quad \tilde{N}_{cj} \right] = \left[Q_{jj} \quad Q_j \right] E_j ,$$

$$Q_{jj} \in \mathrm{H}^{n_{ij} \times n_{ij}}, \ Q_j \in \mathrm{H}^{n_{ij} \times n_{oj}}, \text{ such that } T_m \text{ is H–unimodular } \Big\} , \tag{4.5.13}$$

and equivalently, by

$$S_d(P) = \left[C_1 \ \cdots \ C_m \right] = diag \left[N_{c1} D_{c1}^{-1} \ \cdots \ N_{cm} D_{cm}^{-1} \right] \ \Big|$$

$$\text{for } j = 1, \cdots, m, \begin{bmatrix} -N_{cj} \\ D_{cj} \end{bmatrix} = E_j^{-1} \begin{bmatrix} -\hat{Q}_j \\ \hat{Q}_{jj} \end{bmatrix},$$

$$\hat{Q}_{jj} \in \mathrm{H}^{n_{oj} \times n_{oj}}, \ \hat{Q}_j \in \mathrm{H}^{n_{ij} \times n_{oj}}, \text{ such that } \hat{T}_m \text{ is H–unimodular } \Big\} , \tag{4.5.14}$$

where the matrix T_m is defined in equation (4.5.11) and $\hat{T}_m$ is defined similarly. □

Comment 4.5.6. (The rational functions case)

Let the principal ideal domain H be the ring R_U as in Section 4.4. The main results of Section 4.4 can be extended to the m-channel decentralized control system $\mathrm{S}(P, C_d)_m$ as follows:

The plant P is said to have a *decentralized fixed-eigenvalue* at $s_o \in \bar{\mathrm{U}}$ with respect to $K_d = diag \left[K_1 \ \cdots \ K_m \right]$ iff $s_o \in \bar{\mathrm{U}}$ is a pole of the closed-loop I/O map of the system $\mathrm{S}(P, K_d)$ *for all* $K_d \in \left\{ diag \left[K_1 \ \cdots \ K_m \right] \ \middle| \ K_j \in \mathbb{R}^{n_{ij} \times n_{oj}} \right\}$.

Equivalently, $s_o \in \bar{\mathcal{U}}$ is a decentralized fixed-eigenvalue if and only if

$$\det \begin{bmatrix} D_{p1}(s_o) + K_1 N_{p1}(s_o) \\ \vdots \\ D_{pm}(s_o) + K_m N_{pm}(s_o) \end{bmatrix}$$

$$\sim \det \begin{bmatrix} \tilde{D}_{p1}(s_o) + \tilde{N}_{p1}(s_o) K_1 & \cdots & \tilde{D}_{pm}(s_o) + \tilde{N}_{pm}(s_o) K_m \end{bmatrix} = 0$$

$$\text{for } all \quad K_j \in \mathbb{R}^{n_{ij} \times n_{oj}} \quad , \quad j = 1 , \cdots , m \ . \qquad (4.5.15)$$

To extend Theorem 4.4.4 to the m–channel system $\mathbf{S}(P, C_d)_m$, let P satisfy Assumption 4.5.1 (i); furthermore, let $P \in \mathfrak{m}(\mathbb{R}_{sp}(s))$; then the following six conditions are equivalent:

(i) There exists a decentralized $\mathbf{H}$–stabilizing compensator C_d for P in the system $\mathbf{S}(P, C_d)_m$;

(ii) Any r.c.f.r. (N_p, D_p) of P, partitioned as in equation (4.5.1), satisfies condition (4.5.9) of Theorem 4.5.4, where, for $j = 1, \cdots, m$, $E_j \in \mathbf{R_u}^{(n_{ij}+n_{oj}) \times (n_{ij}+n_{oj})}$ is $\mathbf{R_u}$–unimodular, $R \in \mathbf{R_u}^{n_i \times n_i}$ is $\mathbf{R_u}$–unimodular and $L \in \mathbf{R_u}^{n_o \times n_o}$ is $\mathbf{R_u}$–unimodular; the matrices $W_{jk} \in \mathbf{R_u}^{n_{oj} \times n_{ik}}$ are $\mathbf{R_u}$–stable for $k = 1, \cdots, m$;

(iii) Any l.c.f.r. $(\tilde{D}_p, \tilde{N}_p)$ of P, partitioned as in equation (4.5.2), satisfies condition (4.5.10) of Theorem 4.5.4, where, for $j = 1, \cdots, m$, $E_j \in \mathbf{R_u}^{(n_{ij}+n_{oj}) \times (n_{ij}+n_{oj})}$ is $\mathbf{R_u}$–unimodular, $R \in \mathbf{R_u}^{n_i \times n_i}$ is $\mathbf{R_u}$–unimodular and $L \in \mathbf{R_u}^{n_o \times n_o}$ is $\mathbf{R_u}$–unimodular; the matrices $W_{jk} \in \mathbf{R_u}^{n_{oj} \times n_{ik}}$ are $\mathbf{R_u}$–stable for $k = 1, \cdots, m$.

(iv) For $k = 1, \cdots, m-1$, for all nonempty subsets $A = \{ \alpha_1, \cdots, \alpha_k \}$ of $\{ 1, \cdots, m \}$, any r.c.f.r. (N_p, D_p) of P, partitioned as in equation (4.5.1), satisfies the rank condition (4.5.16) below:

$$rank \begin{bmatrix} D_{p\alpha_1}(s) \\ N_{p\alpha_1}(s) \\ \vdots \\ D_{p\alpha_m}(s) \\ N_{p\alpha_m}(s) \end{bmatrix} \geq \sum_{\alpha_j \in A} n_{i\alpha_j} \ , \quad \text{for all } s \in \bar{\mathbf{u}} \ ; \qquad (4.5.16)$$

(v) For $k = 1, \cdots, m-1$, for all nonempty subsets $A = \{ \alpha_1, \cdots, \alpha_k \}$ of $\{ 1, \cdots, m \}$, any l.c.f.r. $(\tilde{D}_p, \tilde{N}_p)$ of P, partitioned as in equation (4.5.2), satisfies the rank condition (4.5.17) below:

$$rank \left[-\tilde{N}_{p\alpha_1}(s) \ \ \tilde{D}_{p\alpha_1}(s) \ \ \cdots \ \ -\tilde{N}_{p\alpha_k}(s) \ \ \tilde{D}_{p\alpha_k}(s) \right] \geq \sum_{\alpha_j \in A} n_{i\alpha_j} \ ,$$

$$\text{for all } s \in \bar{\mathbf{u}} \ ; \qquad (4.5.17)$$

(vi) The plant P has no decentralized fixed-eigenvalues in $\bar{\mathbf{u}}$.

By Corollary 4.4.5, the six equivalent conditions above are equivalent to condition (vii) below:

(vii) For $k = 1, \cdots, m-1$, for all partitions of the set $\{ 1, \cdots, m \}$ into two (disjoint) subsets $\{ \alpha_1, \cdots, \alpha_k \}$ and $\{ \alpha_{k+1}, \cdots, \alpha_m \}$, any b.c.f.r. (N_{pr}, D, N_{pl}, G) of P, partitioned as in equation (4.5.3), satisfies the rank condition (4.5.18) below:

$$rank \begin{bmatrix} D(s) & -N_{pl\alpha_{k+1}}(s) & \cdots & -N_{pl\alpha_m}(s) \\ N_{pr\alpha_1}(s) & G_{\alpha_1\alpha_{k+1}}(s) & \cdots & G_{\alpha_1\alpha_m}(s) \\ \vdots & \vdots & & \vdots \\ N_{pr\alpha_k}(s) & G_{\alpha_k\alpha_{k+1}}(s) & \cdots & G_{\alpha_k\alpha_m}(s) \end{bmatrix} \geq n \ , \quad \text{for all } s \in \bar{\mathbf{u}} \ .$$

(4.5.18)

Condition (4.5.18) can also be written in the state-space setting as in Remark 4.4.6 by setting $D = \left[(s+a)^{-1} (s I_n - \bar{A}) \right]$, $N_{pr\alpha_j} = \bar{C}_{\alpha_j}$, $N_{pl\alpha_j} = \bar{B}_{\alpha_j}$.

Conditions (4.5.16) and (4.5.17) need not be checked for the entire set $\{ 1, \cdots, m \}$ but only for (proper) subsets of it because since (N_p, D_p) is r.c. and $(\tilde{D}_p, \tilde{N}_p)$ is l.c., Lemma 2.6.1 implies that $rank \begin{bmatrix} D_p(s) \\ N_p(s) \end{bmatrix} = n_i$ and $rank \begin{bmatrix} \tilde{N}_p(s) & \tilde{D}_p(s) \end{bmatrix} = n_i$, for all $s \in \bar{\mathcal{U}}$. Similarly, condition (4.5.18) needs to be checked for all disjoint pairs of subsets neither one of which is all of $\{ 1, \cdots, m \}$; condition (4.5.18) is automatically satisfied if either one of the two subsets were all of $\{ 1, \cdots, m \}$ since (N_{pr}, D, N_{pl}) is a bicoprime triple. □

4.5.7. Achievable input-output maps of $\mathbf{S}(P, C_d)_m$

For a simpler representation than we would obtain for $H_{\overline{yu}}$, we look at the I/O maps of the H–stabilized system $\mathbf{S}(P, C_d)_m$ in a slightly different order.

Let $H_{y'yuu'} : \begin{bmatrix} u_1 \\ u_1' \\ u_2 \\ u_2' \\ \vdots \\ u_m \\ u_m' \end{bmatrix} \mapsto \begin{bmatrix} y_1' \\ y_1 \\ y_2' \\ y_2 \\ \vdots \\ y_m' \\ y_m \end{bmatrix}$. Throughout this section, let P satisfy Assumption 4.5.1 (i); furthermore, let $P \in \mathfrak{m}(\mathbf{G}_s)$; let any r.c.f.r. (N_p, D_p) of P satisfy condition (4.5.9) of Theorem 4.5.4.

The set

$$\mathcal{A}_d(P) := \{ H_{y'yuu'} \mid C_d \text{ H–stabilizes } P \}$$

is called *the set of all achievable I/O maps* of the unity-feedback system $\mathbf{S}(P, C_d)_m$.

By Theorem 4.5.5, $\mathcal{A}_d(P) = \{ H_{y'yuu'} \mid C_d \in \mathcal{S}_d(P) \}$, where $\mathcal{S}_d(P)$ is the set of all decentralized H–stabilizing compensators given by the equivalent representations (4.5.13) and (4.5.14).

Since the set $\mathcal{S}_d(P)$ is a subset of the set $\mathcal{S}(P)$ of all H–stabilizing compensators for the full-feedback system $\mathbf{S}(P, C)$, the set $\mathbf{A}_d(P)$ is also a subset of the set $\mathbf{A}(P)$ of all achievable I/O maps of $\mathbf{S}(P, C)$. The set $\mathbf{A}_d(P)$ is obtained from equations (4.5.6)-(4.5.7) by substituting for (N_p, D_p) from equation (4.5.9) and for $(\tilde{D}_c, \tilde{N}_c)$ from (4.5.13); the matrix T_m in equation (4.5.19) below is the $\mathbf{H}$–unimodular matrix defined in equation (4.5.11):

$$\mathbf{A}_d(P) = \left\{ H_{y'yuu'} = \begin{bmatrix} E_1^{-1} \begin{bmatrix} I_{n_{i1}} & 0 & 0 & \cdots & 0 \\ 0 & W_{12} & W_{13} & \cdots & W_{1m} \end{bmatrix} \\ \\ E_2^{-1} \begin{bmatrix} 0 & I_{n_{i2}} & 0 & \cdots & 0 \\ W_{21} & 0 & W_{23} & \cdots & W_{2m} \end{bmatrix} \\ \vdots \\ E_m^{-1} \begin{bmatrix} 0 & 0 & 0 & \cdots & I_{n_{im}} \\ W_{m1} & W_{m2} & W_{m3} & \cdots & 0 \end{bmatrix} \end{bmatrix} T_m^{-1} \; . \right.$$

$$\begin{bmatrix} \begin{bmatrix} Q_{11} & Q_1 \end{bmatrix} E_1 & 0 & \cdots & 0 \\ 0 & \begin{bmatrix} Q_{22} & Q_2 \end{bmatrix} E_2 & \cdots & 0 \\ \vdots & \vdots & & \vdots \\ 0 & 0 & \cdots & \begin{bmatrix} Q_{mm} & Q_m \end{bmatrix} E_m \end{bmatrix} \Bigg|$$

$$\text{for } j = 1, \cdots, m, \; Q_{jj} \in \mathbf{H}^{n_{ij} \times n_{ij}}, \; Q_j \in \mathbf{H}^{n_{ij} \times n_{oj}}$$

$$\text{are such that } T_m \text{ is } \mathbf{H}\text{–unimodular} \left. \right\} . \qquad (4.5.19)$$

Since T_m depends on the (matrix-) parameters Q_{jj} and Q_j, the parametrization in (4.5.19) of all achievable I/O maps in $\mathbf{S}(P, C_d)_m$ is not affine in these (matrix-) parameters.

REFERENCES

[Åst.1] K. J. Åström, "Robustness of a design method based on assignment of poles and zeros," *IEEE Transactions on Automatic Control*, vol. AC-25, pp. 588-591, 1980.

[And.1] B. D. O. Anderson, D. J. Clements, "Algebraic characterization of fixed modes in decentralized control," *Automatica*, vol. 17, pp. 703-712, 1981.

[And.2] B. D. O. Anderson, "Transfer function matrix description of decentralized fixed modes," *IEEE Transactions on Automatic Control*, vol. AC-27, no. 6, pp. 1176-1182, 1982.

[Bha.1] A. Bhaya, "Issues in the robust control of large flexible structures," *Ph.D. Dissertation,* University of California, Berkeley, 1986.

[Blo.1] H. Blomberg, R. Ylinen, *Algebraic Theory for Multivariable Linear Systems*, Academic Press, 1983.

[Bou.1] B. Bourbaki, *Commutative Algebra*, Addison-Wesley, 1970.

[Bra.1] F. M. Brash, Jr., J. B. Pearson, "Pole placement using dynamic compensators," *IEEE Transactions on Automatic Control*, vol. AC-15, pp. 34-43, 1970.

[Cal.1] F. M. Callier, C. A. Desoer, "An algebra of transfer functions of distributed linear time-invariant systems," *IEEE Transactions on Circuits and Systems*, vol. CAS-25, pp. 651-663, September 1978.

[Cal.2] F. M. Callier, C. A. Desoer, "Stabilization, tracking, and disturbance rejection in multivariable control systems," *Annales de la Société Scientifique de Bruxelles*, T. 94, I, pp. 7-51, 1980.

[Cal.3] F. M. Callier, C. A. Desoer, *Multivariable Feedback Systems*, Springer-Verlag, 1982.

[Che.1] L. Cheng, J. B. Pearson, "Frequency domain synthesis of multivariable linear regulators," *IEEE Transactions on Automatic Control*, vol. AC-26, pp. 194-202, February 1981.

[Chen 1] M.J. Chen, C. A. Desoer, "Necessary and sufficient condition for robust stability of distributed feedback systems," *International Journal of Control*, vol. 35, no. 2, pp. 255-267, 1982.

[Coh.1] P. M. Cohn, *Algebra*, vol. 2, John Wiley, New York, 1977.

[Cor.1] J. P. Corfmat, A. S. Morse, "Decentralized control of linear multivariable systems," *Automatica*, vol. 12, pp. 479-495, 1976.

[Dat.1] K. B. Datta, M. L. J. Hautus, "Decoupling of multivariable control systems over unique factorization domains," *SIAM Journal of Control and Optimization*, vol. 22, no.1, pp. 28-39, 1984.

[Dav.1] E. J. Davison, S. H. Wang, "A characterization of fixed modes in terms of transmission zeros," *IEEE Transactions on Automatic Control*, vol. AC-30, no.1, pp. 81-82, 1985.

[Dav.2] E. J. Davison, T. N. Chang, "Decentralized stabilization and pole assignment for general improper systems," *Proc. American Control Conference*, pp. 1669-1675, 1987.

[Des.1] C. A. Desoer, R. W. Liu, J. Murray, R. Saeks, "Feedback system design: The fractional representation approach to analysis and synthesis," *IEEE Transactions on Automatic Control*, vol. AC-25, pp. 399-412, 1980.

[Des.2] C. A. Desoer, M. J. Chen, "Design of multivariable feedback systems with stable plant," *IEEE Transactions on Automatic Control*, vol. AC-26, pp. 408-415, April 1981.

[Des.3] C. A. Desoer, C. L. Gustafson, "Algebraic theory of linear multivariable feedback systems," *IEEE Transactions on Automatic Control*, vol. AC-29, pp. 909-917, October 1984.

[Des.4] C. A. Desoer, A. N. Gündeş , "Decoupling linear multivariable plants by dynamic output feedback," *IEEE Transactions on Automatic Control*, vol. AC-31, pp. 744-750, August 1986.

[Des.5] C. A. Desoer, A. N. Gündeş , "Algebraic theory of linear time-invariant feedback systems with two-input two-output plant and compensator," *International Journal of Control,* vol. 47, pp. 33-51, January 1988.

[Des.6] C. A. Desoer, A. N. Gündeş , "Algebraic theory of linear feedback systems with full and decentralized compensators," *Electronics Research Lab. Memorandum* No. M88/1, University of California, Berkeley, January 1988.

[Des.7] C. A. Desoer, A. N. Gündeş , "Bicoprime factorizations of the plant and their relation to right- and left-coprime factorizations," *IEEE Transactions on Automatic Control*, vol. 33, pp. 672-676, July 1988.

[Dio.1] J. M. Dion, C. Commault, "On linear dynamic state feedback decoupling," *Proc. 24th Conference on Decision and Control*, pp. 183-188, Dec. 1985.

[Doy.1] J. Doyle, ONR/Honeywell Workshop lecture notes, October 1984.

[Fei.1] A. Feintuch, R. Saeks, *System Theory: A Hilbert Space Approach*, Academic Press, 1982.

[Fes.1] P. S. Fessas, "Decentralized control of linear dynamical systems via polynomial matrix methods," *International Journal of Control*, vol. 30, no. 2, pp. 259-276, 1979.

[Güç .1] A. N. Güçlü, A. B. Özgüler, "Diagonal stabilization of linear multivariable systems," *International Journal of Control*, vol. 43, pp. 965-980, 1986.

[Gus.1] C. L. Gustafson, C. A. Desoer, "Controller design for linear multivariable feedback systems with stable plants, using optimization with inequality constraints," *International Journal of Control*, vol. 37, pp. 881-907, 1983.

[Ham.1] J. Hammer, P. P. Khargonekar, "Decoupling of linear systems by dynamical output feedback," *Math. Systems Theory*, vol. 17, no. 2, pp. 135-157, 1984.

[Hor. 1] I. M. Horowitz, *Synthesis of Feedback Systems*, Academic Press, 1963.

[Jac. 1] *Basic Algebra*, 2nd ed. New York: W. H. Freeman, 1985.

[Kab. 1] M. G. Kabuli, "Factorization approach to linear feedback systems," *Ph.D. Dissertation,* University of California, Berkeley, 1989.

[Kai. 1] T. Kailath, *Linear Systems*, Prentice Hall, 1980.

[Kha. 1] P. P. Khargonekar and E. D. Sontag, "On the Relation Between Stable Matrix Fraction Factorizations and Regulable Realizations of Linear Systems Over Rings," *IEEE Transactions on Automatic Control*, vol. 27, no. 3, pp. 627-638, June 1982.

[Lan.1] S. Lang, *Algebra*, Addison-Wesley, 1971.

[Mac.1] S. MacLane, G. Birkhoff, *Algebra*, 2nd ed., Collier Macmillan, 1979.

[Man.1] V. Manousiouthakis, "On the parametrization of all decentralized stabilizing controllers," *Proc. American Control Conference*, pp. 2108-2111, 1989.

[Mor.1] M. Morari, E. Zafiriou, *Robust Process Control*, Prentice Hall, 1989.

[Net.1] C. N. Nett, "Algebraic aspects of linear control system stability," *IEEE Transactions on Automatic Control*, vol. AC-31, pp. 941-949, 1986.

[Net.2] C. N. Nett, C. A. Jacobson, M. J. Balas, "A Connection Between State-Space and Doubly Coprime Fractional Representations," *IEEE Transactions on Automatic Control,* vol. 29, pp. 831-832, September 1984.

[Ohm 1] D. Y. Ohm, J. W. Howze, S. P. Bhattacharyya, "Structural synthesis of multivariable controllers," *Automatica*, vol. 21, no. 1, pp. 35-55, 1985.

[Per.1] L. Pernebo, "An algebraic theory for the design of controllers for linear multivariable feedback systems," *IEEE Transactions on Automatic Control*, vol. AC-26, pp. 171-194, February 1981.

[Ros.1] H. H. Rosenbrock, "*State-space and Multivariable Theory*", John Wiley, 1980.

[Sae.1] R. Saeks, J. Murray, "Fractional representation, algebraic geometry and the simultaneous stabilization problem," *IEEE Transactions on Automatic Control*, vol. AC-27, pp. 895-904, August 1982.

[Sal.1] S. E. Salcudean, "Algorithms for optimal design of feedback compensators," *Ph.D. Dissertation,* University of California, Berkeley, 1986.

[Sig.1] L. E. Sigler, *Algebra*, Springer-Verlag, 1976.

[Tar.1] M. Tarokh, "Fixed modes in multivariable systems using constrained controllers," *Automatica*, vol. 21, no. 4, pp. 495-497, 1985.

[Vid.1] M. Vidyasagar, *Control System Synthesis: A Factorization Approach*, Cambridge, MA: MIT Press, 1985.

[Vid.2] M. Vidyasagar, H. Schneider, B. Francis, "Algebraic and topological aspects of stabilization," *IEEE Transactions on Automatic Control*, vol. AC-27, pp. 880-894, 1982.

[Vid.3] M. Vidyasagar, N. Viswanadham, "Algebraic characterization of decentralized fixed modes and pole assignment," Report 82-06, University of Waterloo, 1982.

[Vid.4] M. Vidyasagar, N. Viswanadham, "Construction of inverses with prescribed zero minors and applications to decentralized stabilization," *Linear Algebra and Its Applications*, vol. 83, pp. 103-105, 1986.

[Wan.1] S. H. Wang, E. J. Davison, "On the stabilization of decentralized control systems," *IEEE Transactions on Automatic Control*, vol. AC-18, pp. 473-478, 1973.

[Xie 1] X. Xie, Y. Yang, "Frequency domain characterization of decentralized fixed modes," *IEEE Transactions on Automatic Control*, vol. AC-31, pp. 952-954, 1986.

[You.1] D.C. Youla, H. A. Jabr, J. J. Bongiorno, Jr., "Modern Wiener-Hopf design of optimal controllers, Part II: The multivariable case," *IEEE Transactions on Automatic Control*, vol. AC-21, pp. 319-338, 1976.

[Zam.1] G. Zames, "Feedback and optimal sensitivity: Model reference transformations, multiplicative seminorms and approximate inverses," *IEEE Transactions on Automatic Control*, vol. AC-26, pp. 301-320, April 1981.

[Zam.2] G. Zames, D. Bensoussan, "Multivariable feedback, sensitivity and decentralized control," *IEEE Transactions on Automatic Control*, vol. AC-28, pp. 1030-1034, November 1983.

SYMBOLS

I/O	input-output
MIMO	multiinput-multioutput
a := b	a is defined as b
$\mathbb{R}$	real numbers
$\mathbb{C}$	complex numbers
$\mathbb{C}_+$	complex numbers with nonnegative real part
$\mathbb{Z}$	integer numbers
$\mathbb{Z}_+$	nonnegative integer numbers
I_n	nxn identity matrix
det$A$	the determinant of matrix A
H	principal ring
J	group of units of H
I	a multiplicative subset of H
G	ring of fractions of H associated with I
G_s	Jacobson radical of G
$\mathfrak{m}$(H)	the set of matrices with entries in H .
$\mathfrak{u}$	a closed subset of $\mathbb{C}_+$
$\bar{\mathfrak{u}}$	$\mathfrak{u} \cup \{\infty\}$
$R_{\mathfrak{u}}$	ring of proper scalar rational functions which are analytic in $\mathfrak{u}$
$\mathbb{R}_p(s)$	ring of proper scalar rational functions with real coefficients
$\mathbb{R}_{sp}(s)$	set of strictly proper scalar rational functions with real coefficients
r.c. (l.c.)	right-coprime (left-coprime)
r.f.r. (l.f.r.)	right-fraction (left-fraction) representation
r.c.f.r. (l.c.f.r.)	right-coprime-fraction (left-coprime-fraction) representation
b.c. (b.c.f.r.)	bicoprime (bicoprime-fraction representation)
g.c.d.	greatest-common-divisor
l.c.m.	least-common-multiple
$S(P, C)$	the unity-feedback system
$\Sigma(\hat{P}, \hat{C})$	the general feedback system in which the plant and the compensator each have two (vector-) inputs and two (vector-) outputs
$S(P, C_d)$	the two-channel decentralized feedback system
$S(P, K_d)$	the two-channel decentralized constant feedback system
$S(P, C_d)_m$	the m-channel decentralized feedback system

INDEX

Lecture Notes in Control and Information Sciences

Edited by M. Thoma and A. Wyner

Lecture Notes in Control and Information Sciences

Edited by M. Thoma and A. Wyner

Vol. 117: K.J. Hunt
Stochastic Optimal Control Theory
with Application in Self-Tuning Control
X, 308 pages, 1989.

Vol. 118: L. Dai
Singular Control Systems
IX, 332 pages, 1989

Vol. 119: T. Başar, P. Bernhard
Differential Games and Applications
VII, 201 pages, 1989

Vol. 120: L. Trave, A. Titli, A. M. Tarras
Large Scale Systems:
Decentralization, Structure Constraints
and Fixed Modes
XIV, 384 pages, 1989

Vol. 121: A. Blaquière (Editor)
Modeling and Control of Systems
in Engineering, Quantum Mechanics,
Economics and Biosciences
Proceedings of the Bellman Continuum
Workshop 1988, June 13–14, Sophia Antipolis, France
XXVI, 519 pages, 1989

Vol. 122: J. Descusse, M. Fliess, A. Isidori,
D. Leborgne (Eds.)
New Trends in Nonlinear Control Theory
Proceedings of an International
Conference on Nonlinear Systems,
Nantes, France, June 13–17, 1988
VIII, 528 pages, 1989

Vol. 123: C. W. de Silva, A. G. J. MacFarlane
Knowledge-Based Control with
Application to Robots
X, 196 pages, 1989

Vol. 124: A. A. Bahnasawi, M. S. Mahmoud
Control of Partially-Known
Dynamical Systems
XI, 228 pages, 1989

Vol. 125: J. Simon (Ed.)
Control of Boundaries and Stabilization
Proceedings of the IFIP WG 7.2 Conference
Clermont Ferrand, France, June 20–23, 1988
IX, 266 pages, 1989

Vol. 126: N. Christopeit, K. Helmes
M. Kohlmann (Eds.)
Stochastic Differential Systems
Proceedings of the 4th Bad Honnef Conference
June 20–24, 1988
IX, 342 pages, 1989

Vol.127: C. Heij
Deterministic Identification
of Dynamical Systems
VI, 292 pages, 1989

Vol. 128: G. Einarsson, T. Ericson,
I. Ingemarsson, R. Johannesson,
K. Zigangirov, C.-E. Sundberg
Topics in Coding Theory
VII, 176 pages, 1989

Vol. 129: W. A.Porter, S. C. Kak (Eds.)
Advances in Communications and
Signal Processing
VI, 376 pages, 1989.

Vol. 130: W. A. Porter, S. C. Kak,
J. L. Aravena (Eds.)
Advances in Computing and Control
VI, 367 pages, 1989

Vol. 131: S. M. Joshi
Control of Large Flexible Space Structures
IX, 196 pages, 1989.

Vol. 132: W.-Y. Ng
Interactive Multi-Objective Programming
as a Framework for Computer-Aided Control
System Design
XV, 182 pages, 1989.

Vol. 133: R. P. Leland
Stochastic Models for Laser Propagation
in Atmospheric Turbulence
VII, 145 pages, 1989.

Vol. 134: X. J. Zhang
Auxiliary Signal Design in Fault
Detection and Diagnosis
XII, 213 pages, 1989

Vol. 135: H. Nijmeijer, J. M. Schumacher (Eds.)
Three Decades of Mathematical System Theory
A Collection of Surveys at the Occasion of the
50th Birthday of Jan C. Willems
VI, 562 pages, 1989

Vol. 136: J. Zabczyk (Ed.)
Stochastic Systems and Optimization
Proceedings of the 6th IFIP WG 7.1
Working Conference,
Warsaw, Poland, September 12–16, 1988
VI, 374 pages. 1989